Handbook of Process Control and Instrumentation (Chemical Engineering)

Handbook of Process Control and Instrumentation (Chemical Engineering)

Edited by Lawrence Daley

CLANRYE
INTERNATIONAL
www.clanryeinternational.com

Clanrye International,
750 Third Avenue, 9th Floor,
New York, NY 10017, USA

ISBN: 978-1-63240-715-3

Cataloging-in-Publication Data

Handbook of process control and instrumentation : chemical engineering / edited by Lawrence Daley.
   p. cm.
Includes bibliographical references and index.
ISBN 978-1-63240-715-3
1. Chemical process control. 2. Process control. 3. Chemical engineering. I. Daley, Lawrence
TP155.75 .H36 2018
660.281 5--dc23

For information on all Clanrye International publications
visit our website at www.clanryeinternational.com

$\mathcal{C}$LANRYE
INTERNATIONAL

Contents

Preface

Chemical processes are a crucial part of chemical engineering. They refer to naturally and artificially changing one or many chemical compounds. Process control of chemical processes is an operation of controlling and supervising the optimal and safe use of chemicals and chemical reactions in factories and plants. The processes generally monitored in industrial plants are hydration, oxidation, nitrification, reduction, catalysis, etc. This book is a compilation of chapters that discuss the most vital concepts in the field of process control and instrumentation with respect to chemical engineering. It includes a detailed explanation of the various concepts and applications of this subject. This textbook will serve as a reference to a broad spectrum of readers.

To facilitate a deeper understanding of the contents of this book a short introduction of every chapter is written below:

Chapter 1- The process that changes chemicals is known as chemical process. This process either occurs naturally or is caused by external forces. It can consist of either a single step or a number of steps which is termed as unit operations. This section will provide an integrated understanding of chemical processes.

Chapter 2- Chemicals can be manufactured on a large scale in chemical plants. They use specific equipment and technology in manufacturing these chemicals. The chapter on unit operations in a chemical plant offers an insightful focus, keeping in mind the complex subject matter.

Chapter 3- Mathematical model is a simulation of a scientific or industrial process. They can be very useful in the process industry when running actual processes are deemed too expensive. The topics discussed in the chapter are of great importance to broaden the existing knowledge on plant-wide modeling.

Chapter 4- Instrumentation is the term used for measuring and indicating physical quantity. Simple examples of instrumentation systems are sensors and the mechanical thermostat. This section discusses the methods of instrumentation in a critical manner providing key analysis to the subject matter.

I owe the completion of this book to the never-ending support of my family, who supported me throughout the project.

Editor

Fundamentals of Chemical Processes

The process that changes chemicals is known as chemical process. This process either occurs naturally or is caused by external forces. It can consist of either a single step or a number of steps which is termed as unit operations. This section will provide an integrated understanding of chemical processes.

Chemical Process

Chemical processes are designed and operated for manufacturing value added chemicals, the value addition providing the economic incentive for the existence of the process. The fiercely competitive business environment constantly drives research and innovation for significantly improving existing process technologies and for developing new technologies to satisfy man's ever growing needs. On the operation side, the processes are operated to meet key production objectives that include process safety, product specifications (production rate and quality) and environmental regulations. These key production objectives must be satisfied even as the process is subjected to disturbances such as changes in the fresh feed composition, variation in the ambient temperature, equipment fouling, sensor noise / bias etc. In other words, the process operation must ensure proper management of the process variability so that the key production objectives are met even in the presence of the process variability. This naturally leads to the idea of proper management of process variability, the task accomplished by a well designed automatic process control system.

Consider the heat exchanger in Figure. Steam is used to heat a process stream to a certain temperature. Due to variations in the process stream flow rate and inlet temperature, the stream outlet temperature varies over a large range. From the process operation perspective, this is unacceptable since the large variation in the process stream temperature disturbs the downstream unit (eg. a reactor). The installation of a temperature controller that manipulates the steam flow rate to hold the outlet stream temperature constant mitigates this problem to a very large extent. This is illustrated in the outlet stream temperature and steam flow rate profiles in Figure. For open loop operation (no temperature control), the temperature varies over a large range while the steam flow rate remains constant. On the other hand, for closed loop operation (with temperature control), the variation in the outlet stream temperature is significantly lower with the steam flow rate showing large variability. The temperature controller thus transforms the variability in the outlet stream temperature to the steam flow rate. This simple example illustrates the action of a control loop as an agent for transformation of process variability.

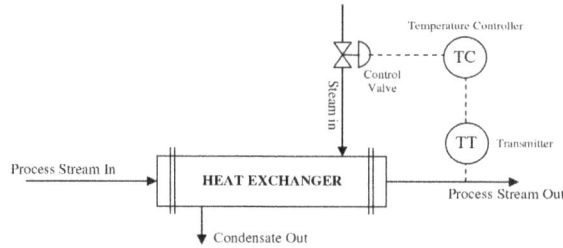

Heat exchanger with temperature controller

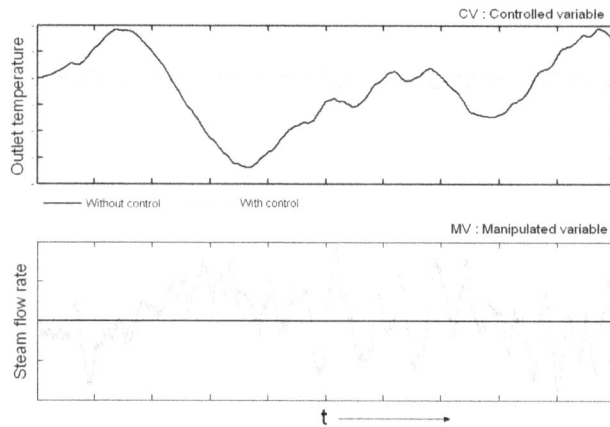

The manipulated (steam flow rate) and controlled variable
(outlet temperature of process stream) with and without control

A chemical process consists of various interconnected units with material and energy recycle. Controlling a process variable by adjusting the flow rate of a process stream necessarily disturbs the downstream / upstream process due to the interconnection. Material and energy recycle can cause the variability to be propagated through the entire plant. Considering the plant-wide propagation / transformation of process variability, the choice of the variables that are controlled (held at / close to their set-points), the corresponding variables that are manipulated and the degree of tightness of control (loose / tight control) are then key determinants of safe and stable process operation. The choice of the controlled and manipulated variables is also sometimes referred to as the control structure. Modern control textbooks provide very little guidance to the practicing engineer on the key issue of control structure selection for individual unit operations and the complete process, choosing instead to focus on the control algorithms and their properties with typical mathematical elegance. How does one go about choosing the most appropriate plant-wide control structure for a given set of production objectives? This work attempts to provide an engineering common sense approach to the practicing engineer for answering this key question.

Given a control system that ensures safe and stable process operation in the face of ever present disturbances, crucial economic variables must be maintained to ensure economically efficient or optimum process operation. Depending on the prevailing economic circumstances, optimality may require process operation at the maximum achievable throughput or lower throughputs. At the optimum steady state, multiple process constraints are usually active such as reactor operation at maximum cooling duty/level/temperature/pressure or column operation at its flooding level etc. How close can the process operate to these constraint limits is intimately tied with the basic plant-wide control strategy implemented. The converse problem is that of designing the regulatory plantwide control system such that the back-off from the constraint limits is the least possible. In this work, we also develop a systematic procedure for designing such an economic plantwide control system.

In a scientific sense, a chemical process is a method or means of somehow changing one or more chemicals or chemical compounds. Such a chemical process can occur by itself or be caused by an outside force, and involves a chemical reaction of some sort. In an "engineering" sense, a chemical process is a method intended to be used in manufacturing or on an industrial scale to change the composition of chemical(s) or material(s), usually using technology similar or related to that used in chemical plants or the chemical industry.

Neither of these definitions is exact in the sense that one can always tell definitively what is a chemical process and what is not; they are practical definitions. There is also significant overlap in these two definition variations. Because of the inexactness of the definition, chemists and other scientists use the term "chemical process" only in a general sense or in the engineering sense. However, in the "process (engineering)" sense, the term "chemical process" is used extensively. The rest of the article will cover the engineering type of chemical process.

Although this type of chemical process may sometimes involve only one step, often multiple steps, referred to as unit operations, are involved. In a plant, each of the unit operations commonly occur in individual vessels or sections of the plant called units. Often, one or more chemical reactions are involved, but other ways of changing chemical (or material) composition may be used, such as mixing or separation processes. The process steps may be sequential in time or sequential in space along a stream of flowing or moving material. For a given amount of a feed (input) material or product (output) material, an expected amount of material can be determined at key steps in the process from empirical data and material balance calculations. These amounts can be scaled up or down to suit the desired capacity or operation of a particular chemical plant built for such a process. More than one chemical plant may use the same chemical process, each plant perhaps at differently scaled capacities. Chemical processes like distillation and crystallization go back to alchemy in Alexandria, Egypt.

Such chemical processes can be illustrated generally as block flow diagrams or in more detail as process flow diagrams. Block flow diagrams show the units as blocks and the streams flowing between them as connecting lines with arrowheads to show direction of flow.

In addition to chemical plants for producing chemicals, chemical processes with similar technology and equipment are also used in oil refining and other refineries, natural gas processing, polymer and pharmaceutical manufacturing, food processing, and water and wastewater treatment.

Unit processing in Chemical Process

Unit processing is the basic processing in chemical engineering. Together with unit operations it forms the main principle of the varied chemical industries. Each genre of unit processing follows the same chemical law much as each genre of unit operations follows the same physical law.

Chemical engineering unit processing consists of the following important processes:

- Oxidation
- Reduction
- Hydrogenation
- Dehydrogenation
- Hydrolysis
- Hydration
- Dehydration

- Halogenation

- Nitrification

- Sulfonation

- Ammoniation

- Alkaline fusion

- Alkylation

- Dealkylation

- Esterification

- Polymerization

- Polycondensation

- Catalysis

Process Integration

Process integration is a term in chemical engineering which has two possible meanings.

1. A holistic approach to process design which emphasizes the unity of the process and considers the interactions between different unit operations from the outset, rather than optimising them separately. This can also be called *integrated process design* or *process synthesis*. El-Halwagi (1997 and 2006) and Smith (2005) describe the approach well. An important first step is often *product design* (Cussler and Moggridge 2003) which develops the specification for the product to fulfil its required purpose.

2. *Pinch analysis*, a technique for designing a process to minimise energy consumption and maximise heat recovery, also known as *heat integration, energy integration* or *pinch technology*. The technique calculates thermodynamically attainable *energy targets* for a given process and identifies how to achieve them. A key insight is the pinch temperature, which is the most constrained point in the process. The most detailed explanation of the techniques is by Linnhoff et al. (1982), Shenoy (1995) and Kemp (2006). This definition reflects the fact that the first major success for process integration was the thermal pinch analysis addressing energy problems and pioneered by Linnhoff and co-workers. Later, other pinch analyses were developed for several applications such as mass-exchange networks (El-Halwagi and Manousiouthakis, 1989), water minimization (Wang and Smith, 1994), and material recycle (El-Halwagi et al., 2003). A very successful extension was "Hydrogen Pinch", which was applied to refinery hydrogen management (Nick Hallale et al., 2002 and 2003). This allowed refiners to minimise the capital and operating costs of hydrogen supply to meet ever stricter environmental regulations and also increase hydrotreater yields.

In the context of chemical engineering, Process Integration can be defined as a holistic approach to process design and optimization, which exploits the interactions between different units in order to employ resources effectively and minimize costs.

Process Integration is not limited to the design of new plants, but it also covers retro-fit design (e.g. new units to be installed in an old plant) and the operation of existing systems. Nick Hallale (2001), in his article in Chemical Engineering Progress provided a state of the art review. He explained that process integration far wider scope and touches every area of process design. Industries are making more money from their raw materials and capital assets while becoming cleaner and more sustainable.

The main advantage of process integration is to consider a system as a whole (i.e. integrated or holistic approach) in order to improve their design and/or operation. In contrast, an analytical approach would attempt to improve or optimize process units separately without necessarily taking advantage of potential interactions among them.

For instance, by using process integration techniques it might be possible to identify that a process can use the heat rejected by another unit and reduce the overall energy consumption, even if the units are not running at optimum conditions on their own. Such an opportunity would be missed with an analytical approach, as it would seek to optimize each unit, and thereafter it wouldn't be possible to re-use the heat internally.

Typically, process integration techniques are employed at the beginning of a project (e.g. a new plant or the improvement of an existing one) to screen out promising options to optimize the design and/or operation of a process plant.

Also it is often employed, in conjunction with simulation and mathematical optimization tools to identify opportunities in order to better integrate a system (new or existing) and reduce capital and/or operating costs.

Most process integration techniques employ Pinch analysis or Pinch Tools to evaluate several processes as a whole system. Therefore, strictly speaking, both concepts are not the same, even if in certain contexts they are used interchangeably. The review by Nick Hallale (2001) explains that in the future, several trends are to be expected in the field. In the future, it seems probable that the boundary between targets and design will be blurred and that these will be based on more structural information regarding the process network. Second, it is likely that we will see a much wider range of applications of process integration. There is still much work to be carried out in the area of separation, not only in complex distillation systems, but also in mixed types of separation systems. This includes processes involving solids, such as flotation and crystallization. The use of process integration techniques for reactor design has seen rapid progress, but is still in its early stages. Third, a new generation of software tools is expected. The emergence of commercial software for process integration is fundamental to its wider application in process design.

Process Design

In chemical engineering, process design is the design of processes for desired physical and/or chemical transformation of materials. Process design is central to chemical engineering, and it can be considered to be the summit of that field, bringing together all of the field's components.

Process design can be the design of new facilities or it can be the modification or expansion of existing facilities. The design starts at a conceptual level and ultimately ends in the form of fabrication and construction plans.

Process design is distinct from equipment design, which is closer in spirit to the design of unit operations. Processes often include many unit operations.

Documentation

Process design documents serve to define the design and they ensure that the design components fit together. They are useful in communicating ideas and plans to other engineers involved with the design, to external regulatory agencies, to equipment vendors and to construction contractors.

In order of increasing detail, process design documents include:

- Block flow diagrams (BFD): Very simple diagrams composed of rectangles and lines indicating major material or energy flows.

- Process flow diagrams (PFD): Typically more complex diagrams of major unit operations as well as flow lines. They usually include a material balance, and sometimes an energy balance, showing typical or design flowrates, stream compositions, and stream and equipment pressures and temperatures.

- Piping and instrumentation diagrams (P&ID): Diagrams showing each and every pipeline with piping class (carbon steel or stainless steel) and pipe size (diameter). They also show valving along with instrument locations and process control schemes.

- Specifications: Written design requirements of all major equipment items.

Process designers also typically write operating manuals on how to start-up, operate and shut-down the process.

Documents are maintained after construction of the process facility for the operating personnel to refer to. The documents also are useful when modifications to the facility are planned.

A primary method of developing the process documents is process flowsheeting.

Design Considerations

There are several considerations that need to be made when designing any chemical process unit. Design conceptualization and considerations can begin once product purities, yields, and throughput rates are all defined.

Objectives that a design may strive to include:

- Throughput rate
- Process yield
- Product purity

Constraints include:

- Capital cost
- Available space
- Safety concerns
- Environmental impact and projected effluents and emissions
- Waste production
- Operating and maintenance costs

Other factors that designers may include are:

- Reliability
- Redundancy
- Flexibility
- Anticipated variability in feedstock and allowable variability in product.

Sources of Design Information

Designers usually do not start from scratch, especially for complex projects. Often the engineers have pilot plant data available or data from full-scale operating facilities. Other sources of information include proprietary design criteria provided by process licensors, published scientific data, laboratory experiments, and input.

Computer Help

The advent of low cost powerful computers has aided complex mathematical simulation of processes, and simulation software is often used by design engineers.

Simulations can identify weaknesses in designs and allow engineers to choose better alternatives.

However, engineers still rely on heuristics, intuition, and experience when designing a process. Human creativity is an element in complex designs.

Process Control

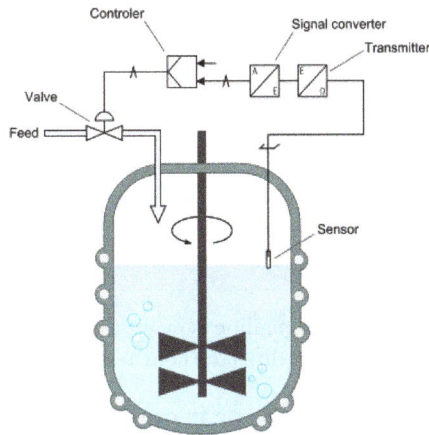

Example of control system of a continuous stirred-tank reactor.

Control panel of a nuclear reactor.

Process control is an engineering discipline that deals with architectures, mechanisms and algorithms for maintaining the output of a specific process within a desired range. For instance, the temperature of a chemical reactor may be controlled to maintain a consistent product output.

Process control is extensively used in industry and enables mass production of consistent products from continuously operated processes such as oil refining, paper manufacturing, chemicals, power plants and many others. Process control enables automation, by which a small staff of operating personnel can operate a complex process from a central control room.

Background

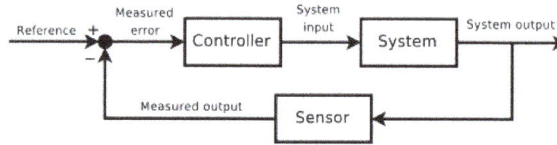

Block diagram of a closed control (Feedback) loop

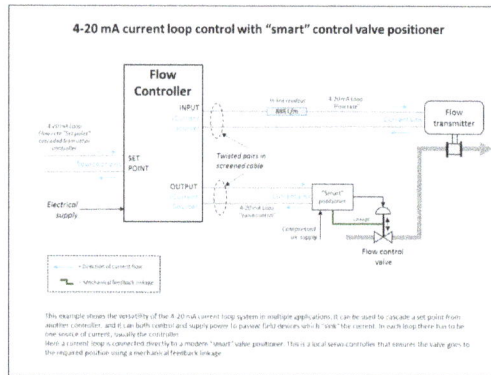

Example of a continuous flow control loop. Signalling is by industry standard 4-20 mA current loops, and a "smart" valve positioner ensures the control valve operates correctly.

Process control may either use feedback or it may be open loop. Control may also be continuous (automobile cruise control) or cause a sequence of discrete events, such as a timer on a lawn sprinkler (on/off) or controls on an elevator (logical sequence).

A thermostat on a heater is an example of control that is on or off. A temperature sensor turns the heat source on if the temperature falls below the set point and turns the heat source off when the set point is reached. There is no measurement of the difference between the set point and the measured temperature (e.g. no error measurement) and no adjustment to the rate at which heat is added other than all or none.

A familiar example of feedback control is cruise control on an automobile. Here speed is the measured variable. The operator (driver) adjusts the desired speed set point (e.g. 100 km/hr) and the controller monitors the speed sensor and compares the measured speed to the set point. Any deviations, such as changes in grade, drag, wind speed or even using a different grade of fuel (for example an ethanol blend) are corrected by the controller making a compensating adjustment to the fuel valve open position, which is the manipulated variable. The controller makes adjustments having information only about the error (magnitude, rate of change or cumulative error) although settings known as *tuning* are used to achieve stable control. The operation of such controllers is the subject of control theory.

A commonly used control device called a programmable logic controller, or a PLC, is used to read a set of digital and analog inputs, apply a set of logic statements, and generate a set of analog and digital outputs.

For example, if an adjustable valve were used to hold level in a tank the logical statements would compare the equivalent pressure at depth setpoint to the pressure reading of a sensor below the normal low liquid level and determine whether more or less valve opening was necessary to keep the level constant. A PLC output would then calculate an incremental amount of change in the valve position. Larger more complex systems can be controlled by process control systems like Distributed Control System (DCS) or SCADA.

Hierarchy of Process Control

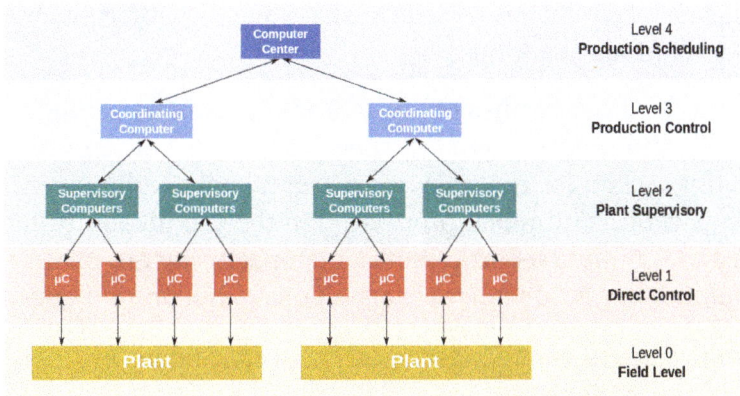

Functional levels of a manufacturing control operation.

The accompanying diagram is a general model which shows functional manufacturing levels in a large process using computerised control.

Referring to the diagram:

- Level 0 contains the field devices such as flow and temperature sensors, and final control elements, such as control valves.

- Level 1 contains the industrialised Input/Output (I/O) modules, and their associated distributed electronic processors.

- Level 2 contains the supervisory computers, which collate information from processor nodes on the system, and provide the operator control screens.

- Level 3 is the production control level, which does not directly control the process, but is concerned with monitoring production and monitoring targets.

- Level 4 is the production scheduling level.

Types of Processes using Process Control

Processes can be characterized as one or more of the following forms:

- Discrete – Found in many manufacturing, motion and packaging applications.

Robotic assembly, such as that found in automotive production, can be characterized as discrete process control. Most discrete manufacturing involves the production of discrete pieces of product, such as metal stamping.

- Batch – Some applications require that specific quantities of raw materials be combined in specific ways for particular durations to produce an intermediate or end result. One example is the production of adhesives and glues, which normally require the mixing of raw materials in a heated vessel for a period of time to form a quantity of end product. Other important examples are the production of food, beverages and medicine. Batch processes are generally used to produce a relatively low to intermediate quantity of product per year (a few pounds to millions of pounds).

- Continuous – Often, a physical system is represented through variables that are smooth and uninterrupted in time. The control of the water temperature in a heating jacket, for example, is an example of continuous process control. Some important continuous processes are the production of fuels, chemicals and plastics. Continuous processes in manufacturing are used to produce very large quantities of product per year (millions to billions of pounds).

Applications having elements of discrete, batch and continuous process control are often called *hybrid* applications.

Examples

- An anti-lock braking system (ABS) is a complex example, consisting of multiple inputs, conditions and outputs.

- Aircraft stability control is a highly complex example using multiple inputs and outputs.

Process Dynamics

Process dynamics refers to the time trajectory of a variable in response to a change in an input to the process. All of us have an inherent appreciation of process dynamics in the sense that the effect of a cause takes time to manifest itself. It thus takes 20 minutes for a pot of rice to cook over a flame, 5-10 minutes for the water in the geyser to heat up sufficiently, years and years of dedicated practice to become an adept musician (or a good engineer, for that matter!) and so on so forth. In each of these examples, a change in the causal variable (flame, electric heating or dedicated practice) results in a change over time in the effected variable (degree of "cookedness" of rice, geyser water temperature or a musician's virtuosity). Process dynamics deals with the systematic characterization of the time response of the effected variable to a change in the causal variable. In process control parlance, the causal variable is referred to as an input variable and the effected variable is referred to as an output variable.

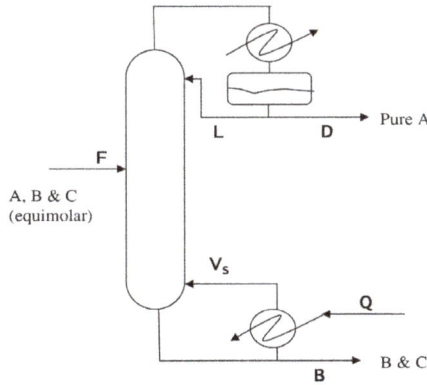

F   : Feed
D   : Distillate
L   : Reflux
V$_s$: Boil-up
B   : Bottoms
Q   : Heat Duty

Schematic of a distillation column

In order to fix ideas in the context of chemical processes, Figure shows the schematic of a simple distillation column. An equimolar ABC feed is separated to recover nearly pure A as the distillate with the bottoms being a BC mixture with trace amounts of A. The fresh feed, reflux and reboil constitute the inputs to the column while the distillate and bottoms flow / composition and the tray composition / temperature profiles constitute the outputs.

Standard Input Changes

To systematically characterize the transient response of an output to a change in the input, the input change is usually standardized to a step change, a pulse change or an impulse change. These standard input changes are depicted in Figure. A step change in the input, the simplest input change pattern, is used in this work to characterize the process dynamics.

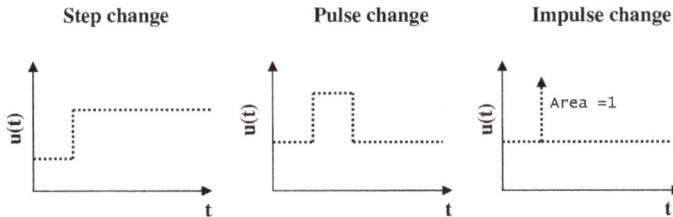

Standard input changes

Basic Response Types

The dynamics of every process are. Even so, the variety of transient responses can be characterized as an appropriate combination of one or more basic response types. These transient responses correspond to the solution of linear ordinary differential equations (ODEs). Linear ODEs can be compactly represented using Laplace transforms. For example consider a second order differential equation

$$\tau^2 \frac{d^2 y(t)}{dt^2} + 2\zeta r \frac{dy(t)}{dt} + y(t) = K_p u(t)$$

where y(t) and u(t) are the process output and input respectively. The Laplace transform representation in the s domain is obtained by replacing the nth order derivative operator by s^n so that for the second order ODE above

$$\tau^2.s^2 y(s) + 2\zeta\tau.s.y(s) + y(s) = K_p u(s)$$

Rearranging, the input-output transfer function becomes

$$G_p = \frac{y(s)}{u(s)} = \frac{K_p}{\tau^2 s^2 + 2\zeta\tau s + 1}$$

The ODEs and corresponding Laplace transform representation is noted in Table.

First Order Lag

The first order lag is the simplest transient response where the output immediately responds to a step change in the input. The ratio of the change in the output to the change in the input is referred to as the process gain, K_p. The time it takes for the output to reach 63.2% of its final value corresponds to the first order time constant τ_p. The output reaches ~95% of its final value in 3 time constants.

Table: Various Differential Equations and their Laplace Transform

Terminology	Differential equation	Laplace Transform $\dfrac{y(s)}{u(s)}$
Gain	$y(t) = K.u(t)$	K
Derivative	$y(t) = \dfrac{du(t)}{dt}$	s
Integrator	$y(t) = \int_0^t u(t).dt$	$\dfrac{1}{s}$
First order lag	$\tau\dfrac{dy(t)}{dt} + y(t) = u(t)$	$\dfrac{1}{\tau s + 1}$
First-order lead	$\tau\dfrac{du(t)}{dt} + u(t) = y(t)$	$\tau s + 1$
Second Order Lag		
Underdamped $\zeta < 1$	$\tau^2\dfrac{d^2 y(t)}{dt^2} + 2\zeta\tau\dfrac{dy(t)}{dt} + y(t) = K_p u(t)$	$\dfrac{K_p}{\tau^2 s^2 + 2\zeta\tau s + 1}$
Critically damped $\zeta = 1$	$\tau^2\dfrac{d^2 y(t)}{dt^2} + 2\tau\dfrac{dy(t)}{dt} + y(t) = K_p u(t)$	$\dfrac{K_p}{(\tau s + 1)^2}$
Overdamped $\zeta > 1$	$\tau_1\tau_2\dfrac{d^2 y(t)}{dt^2} + \zeta(\tau_1 + \tau_2)\dfrac{dy(t)}{dt} + y(t) = K_p u(t)$	$\dfrac{K_p}{(\tau_1 s + 1)(\tau_2 s + 1)}$
Deadtime	$y(t) = u(t - \theta)$	$e^{-\theta s}$
Lead-lag	$\tau_2\dfrac{dy(t)}{dt} + y(t) = \tau_1\dfrac{du(t)}{dt} + u(t)$	$\dfrac{\tau_1 s + 1}{\tau_2 s + 1}$

Higher Order Lags

If the output from a first order lag is input to another first order lag, the latter's output behaves as a second order lag with respect to the input to the first lag. The overall transient response is S shaped with the output not responding immediately to a change in the input. When the time constant of the two lags are different, the response is called an over-damped second order response. The response for the special case where the two time constants are equal is called the critically damped second order response. Higher order systems result as more first order lags are connected in series with the transient response becoming increasingly sluggish.

Second Order Response

Sometimes, a step change in the input causes the output to oscillate before settling at the final steady state. The simplest such response corresponds to a second order under-damped system. The damping coefficient, ζ, can be used to characterize all second order responses – overdamped ($\zeta > 1$), critcally damped ($\zeta = 1$) and underdamped ($\zeta < 1$). The second order response is shown in Figure (b).

To gain an appreciation of the impact of damping coefficient on the transient response, Table reports the ratio of the second overshoot to the first overshoot for different values of ζ. A quarter decay ratio is observed for a damping coefficient of 0.218. Sustained oscillations (decay ratio = 1) are observed for a damping coefficient of 0. As ζ increases to 1, the overshoot in the output disappears.

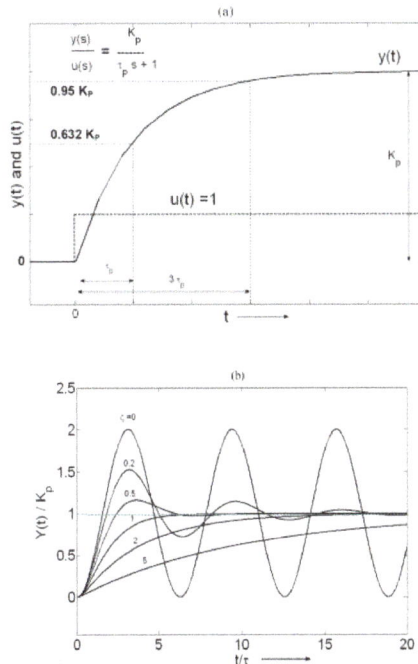

Output response for unit step change to (a) First order & (b) Second order process

Table: Decay Ratio for Various Different Damping Coefficients

Damping Coefficient, ζ	0	0.05	0.1	0.2	0.218	0.4	0.6	1
Decay ratio	1.000	0.730	0.532	0.277	0.250	0.064	0.009	0.000

Other Common Response Types

Other types of responses include the pure integrator, the pure dead-time, and the inverse response. The transient response to a unit step change can be seen in figure and are self explanatory.

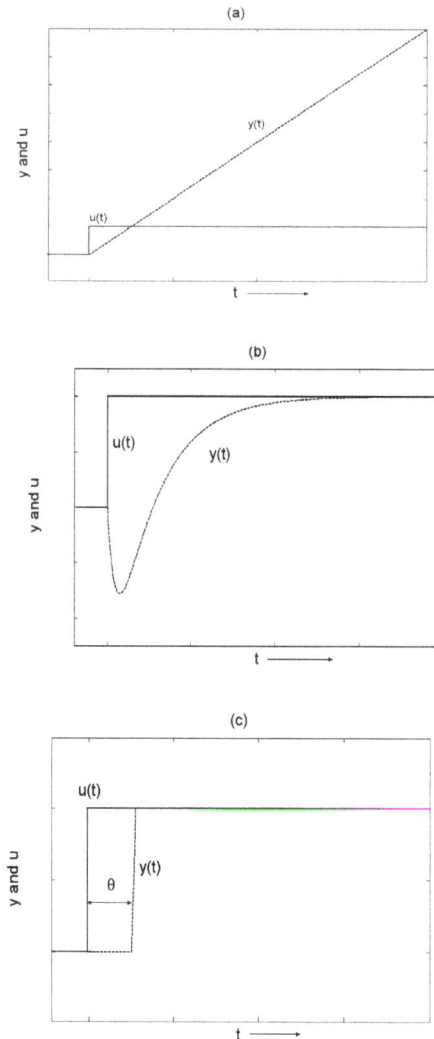

The output response for a unit step change for (a) pure integrator, (b) inverse response and (c) pure dead time process.

The most common example of a pure integrator is the response of the tank level to change in the inlet / outlet feed rate. Unless the inlet and outlet flows are perfectly equal, the tank level is either rising or falling in direct proportion to the mismatch in the flows. The level in a tank is thus non-self regulating with respect to the connected flows. A controller must be used to stabilize all such non-self regulating process variables. Dead time is very common in chemical processing systems and is due to transportation delay. A very common example of the inverse response is the response of the liquid level in a boiler to a change in the heating duty. As the heating duty is increased, the vapour volume entrapped in the liquid increases causing the liquid interface level to rise initially. Over longer duration, the level of course reduces since more liquid is being vaporized. As will be seen later, dead time and inverse response can create control difficulties.

Unstable Systems

Some systems may be inherently unstable. Unstable transient responses are shown in Figure. The unstable response may be non-oscillatory or oscillatory as in the Figure. Reactor temperature runaway is an example of an unstable process. A control system must be used to stabilize an inherently unstable system.

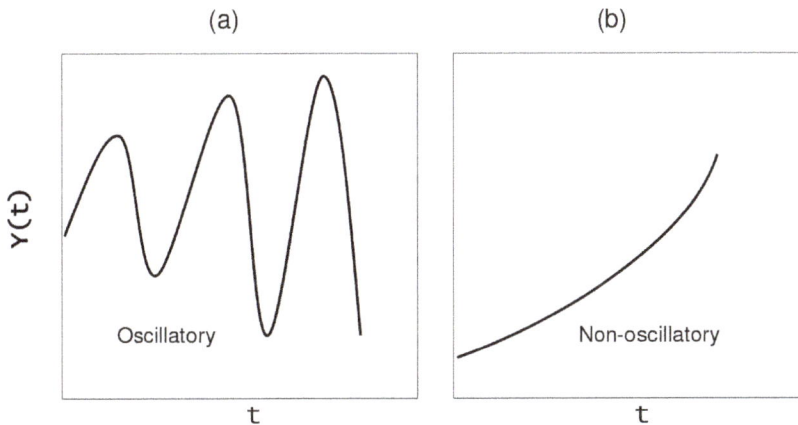

The output response for unstable process. (a) Oscillatory and (b) non-oscillatory

Combination of Basic Responses

Any transient response can be reasonably represented as a combination of the above basic response types. One such combination is the first order lag plus dead time that has been found to represent the transient response of many chemical processing systems very well. The response is illustrated in Figure (a). Another example of such a combination is the inverse response which can be represented by the parallel combination of two first order lags. One of the lags has a small gain and a small time constant (ie a fast response) while the other lag has a gain of larger magnitude and opposite sign with a much larger time constant (i.e. a slow response in the opposite direction). Figure (b) illustrates this concept.

(a)

(b)

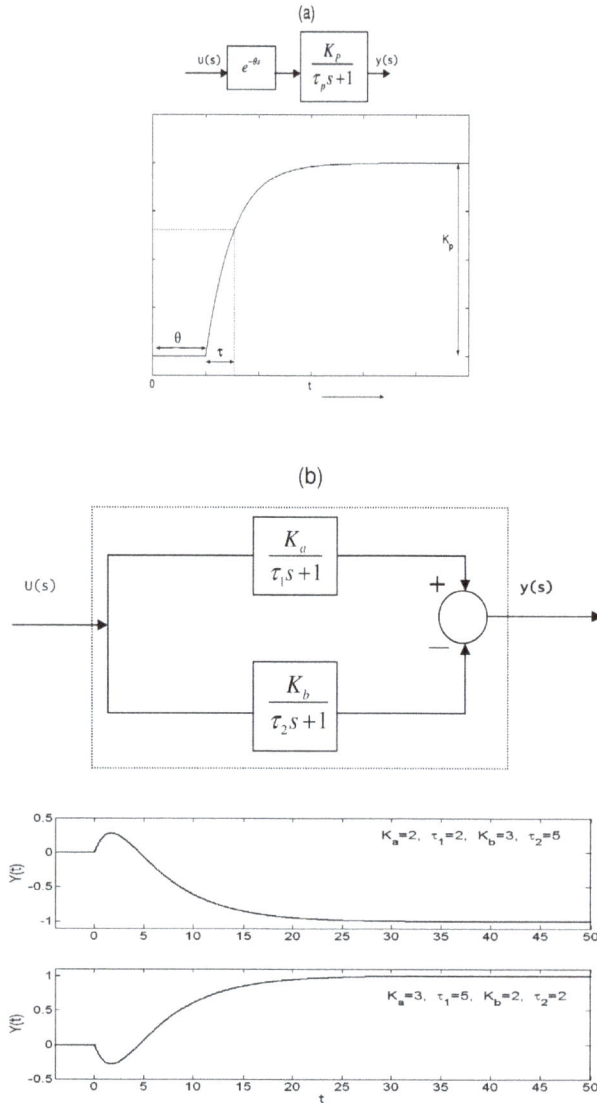

Unit step responses (a) first order plus dead time process (FOPDT)
and (b) Inverse response

Feedback Control

The safe and stable operation of a process requires that key variables be maintained at or close to their design values in the face of disturbances entering the process. For example, it may be necessary to hold a process stream flow rate nearly constant even as the upstream / downstream pressure fluctuates. Similarly the temperature at the inlet to a packed bed reactor must be maintained at its design value to prevent reactor run-away and also ensure the desired conversion to products(s) for varying flow rates of the process stream. Maintaining a process variable at or near a certain value requires a manipulation handle that can be appropriately adjusted. For example, the valve opening can be adjusted to maintain the flow rate through the pipe. Similarly the heating duty of the furnace can be used to heat the process to maintain the reactor inlet stream temperature. This leads to the idea of feedback control where the deviation in the variable to be maintained at / near its design value is used to make appropriate adjustments in the manipulation handle. The variable to be maintained at its design value is referred to as the controlled variable and the adjustment handle is called the manipulated variable. The algorithm / procedure used to quantitatively translate the deviation in the controlled variable to the adjustment in the manipulated variable is known as the control algorithm.

The Feedback Loop and its Components

A feedback control loop is schematically illustrated in figure. Its primary components are the sensor, transducer, transmitter, controller, I/P converter and the final control element. The sensor is the sensing element used to measure the controlled variable (and other important process variables that may not be controlled). Flow, temperature and pressure sensors are routinely used in the process industry. Composition analyzers are used less frequently to measure only key compositions such as the product purity. Most sensors translate a change in the state of the variable to be measured into an equivalent mechanical signal such as the stretching / bending of a Bourdon tube. The mechanical signal needs to be converted into an electrical signal for onward transmission to the control room (or stand- alone controller). This is accomplished by the transducer. For standardization across different manufacturers, the range of the input and output signal from a controller is 4-20 mA. The range corresponds to the sensor / final control element span. The transmitter converts the electrical signal from the transducer to the 4-20 mA range. The transmitter signal is input to the controller. The desired value for the controlled variable, referred to as the set-point, is also input to the controller. The controller output signal is again between 4-20 mA. In the process industry, this electrical signal is converted to an equivalent 3-15 psig pneumatic pressure signal using an I/P converter. The pressure signal (or rather change in the pressure signal) is used to move the final control element to bring about a change in the manipulated variable. In the process industry, almost all final control elements are control valves that adjust the flow rate of a material stream.

The controller subtracts the current value of the controlled variable from its set-point to obtain the error signal as

$$e_t = y^{SP} - y_t$$

where y is the controlled variable. The subscript t refers to the current time. The error signal is input to the control algorithm to determine the change in the manipulated variable (control input) to be implemented. This is schematically illustrated in figure. The most popular control algorithm, namely the PID algorithm is discussed next.

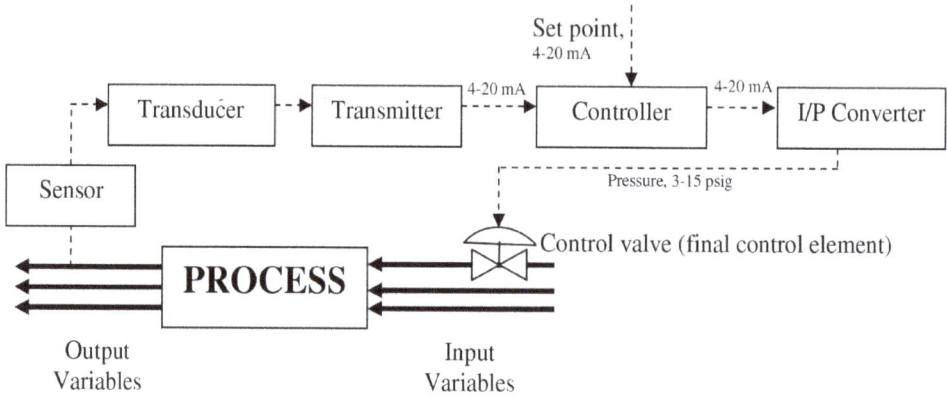

Schematic of a process with feed back control scheme

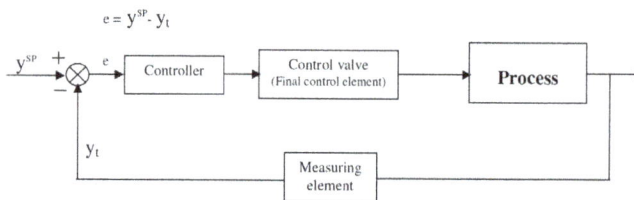

Block diagram of a feed back control

PID Control

The Control Algorithm

Almost all controllers in the process industry use the Proportional Integral Derivative (PID) control algorithm. Even as instrumentation and computation technologies have witnessed a transition from the analog era to the digital revolution, the good old PID control algorithm remains the most widely used algorithm, not withstanding the on-slaught of advanced model predictive control algorithms. The positional form of the algorithm states that

$$u_t = K_c\left(e_t + \frac{1}{\tau_I}\int_o^t e(t)\,dt + \tau_D\frac{de_t}{dt}\right) + bias$$

where u is the controller output (input to the process), e is the error in the controlled variable, and K_c, τ_I and τ_D are controller tuning parameters. The tuning parameters are referred to respectively as the controller gain, reset (or integral time) and derivative time. The bias term in the expression is provided to make the LHS equal the RHS at time t = 0 for proper initialisation. The three terms in the algorithm correspond to Pro-portional, Integral and Derivative action, hence the acronym PID.

The velocity form of the algorithm is more amenable to understanding the effect of each of the P, I and D actions. Differentiating the above equation, we get

$$\frac{du_t}{dt} = K_c \left(\frac{de_t}{dt} + \frac{1}{\tau_I} e_t + \tau_D \frac{d^2 e_t}{dt^2} \right)$$

The controller gain or proportional gain, K_C, determines the fastness of response with larger values resulting in a fast response to deviations from set-point. This can be verified from the first term in the velocity form equation where the rate of change of the control input is directly proportional to the rate of change in the error, K_C being the proportionality constant. The larger the K_C, the larger the change in the control input, the faster the return to set-point.

The integral action is provided to ensure zero offset in the controlled variable. If the controlled variable deviates from its set-point, the controller acts to settle the system at a new steady state. At this new steady state all time derivatives are zero (by definition) implying the LHS in the equation above is zero. The RHS also therefore must be zero which requires that the error term, et, must be zero at the final steady state ($t \rightarrow \infty$). The error term in the velocity form above is due to the integral mode so that integral action moves the control input until the error in the controlled variable is driven to zero i.e. ensures a zero offset. P and D action do not guarantee zero offset as at the final steady state, the LHS and RHS terms corresponding to P and D action are zero. For a P or PD controller with no integral action, the velocity form of the algorithm imposes no restriction on the output error at the final steady state. A non-zero offset thus can and does result sans integral action.

The derivative action causes the controller to "think ahead" and is usually introduced to suppress oscillations from the "seeking behaviour" caused by integral action. In effect, the derivative action puts brakes on the control action as the controlled variable approaches the set-point thus avoiding large oscillations around the set-point. Most controllers in the industry are P or PI controllers and the D action is set to zero. This is because the D action amplifies noise so that the controller input signal must be pre-filtered appropriately to reap the benefit of D action. It is easier to simply turn the D action off and properly tune the controller gain and reset time for the desired control performance.

Controller Tuning

Empirical rules have been developed for tuning PID controllers. These tuning rules are based on the idea of ultimate gain and ultimate period. Figure plots the closed loop response for a unit step change in the set-point of a first order plus dead time process for a P only controller as the controller gain is increased. Notice that as the controller gain is increased, the steady state offset reduces. Also, the response becomes faster. For larger gains the closed loop response is oscillatory. As the gain is increased further, sustained oscillations result. Any further increase in the controller gain results in an unstable system with the oscillations increasing in magnitude with time. The controller gain for which the closed loop response exhibits sustained oscillations corresponds to the transition from a stable to an unstable closed loop response. This controller gain at which the closed loop system borders on instability is referred to as the ultimate gain, K_U. The period of the sustained oscillations is known as the ultimate period, P_U. The empirical tuning rules recommend the controller gain to be a fraction of the ultimate gain and the reset time and derivative time as fractions (multiples) of P_U. Two popular tuning rules are the Zeigler-Nichols and Tyreus-Luyben tuning rules are tabulated in Table. For a given ultimate gain and ultimate period, the controller gain is the least for a PI controller. This is due to the "seeking behaviour" caused by integral action for zero offset. The closed loop system thus goes unstable for a lower controller gain implying that it should be lower. The controller gain is the maximum for a PID controller due to the stabilizing effect of D action. As discussed before, D action is however used rarely in practice due to noise amplification. The PI algorithm is most commonly used in the industry. The tuning rules show that Zeigler-Nichols tuning is more aggressive than the Tyreus-Luyben tuning. Application of the ZN tuning rule can cause process upsets such as a distillation column flooding due to a sudden large increase in the vapour boil-up caused by a controller. The more conservative TL tuning rule is preferred in the process industry for a smooth and bumpless handling of transients avoiding large and sudden changes in the control input.

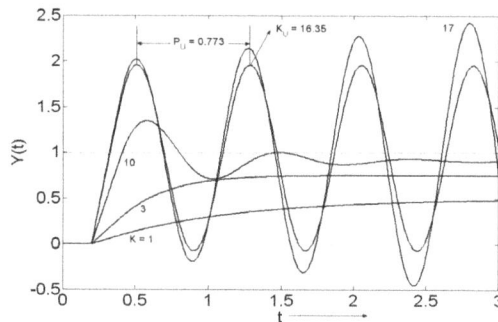

Closed loop response of a first order plus dead time process using
P controller with different controller gains (K).

Table:

	P	PI	PID
Ziegler-Nichols			
K_C	$K_U/2$	$K_U/2.2$	$K_U/1.7$
τ_I	--	$P_U/1.2$	$P_U/2$
τ_D	--	--	$P_U/8$
Tyreus -Luyben			

K_C	--	$K_U/3.2$	Ku/2.2
τ_I		$2.2P_U$	$2.2P_U$
τ_D	--	--	$P_U/6.3$

It is appropriate to highlight that a controller is required to handle two types of changes namely, a change in the output set-point and a change in the measured / unmeasured disturbance into the process. The closed loop response for these is respectively referred to as the servo and the regulator response. A disturbance into a process is also sometimes referred to as a load change. Control systems in the process industry are typically designed for effective load rejection. In contrast, set-point tracking is the primary objective in the design of control systems for aerospace systems such as aeroplanes, rockets and missiles.

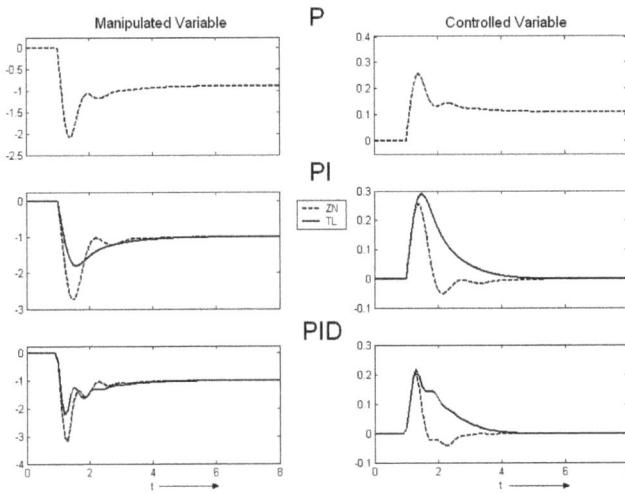

Dynamics of manipulated and controlled variables using P, PI and PID controllers with ZN and TL controller parameters for a unit step change in load. (Regulatory response).

Figure plots the regulator response for a unit step in the load variable with a P, PI and PID controller tuned using the ZN and TL tuning rules for the first order plus dead time process considered earlier. Notice that P only control results in an offset at the final steady state. This offset is larger for TL tuning due to the lower controller gain. The PI and PID regulator responses show no offset at the final steady state due to integral action. Also notice that the aggressive ZN tuning results in a quicker but oscillatory return to the set-point for the PI controller. These oscillations are suppressed by the D action in a PID controller. PID control leads to a faster and smoother return to set-point due to the stabilizing effect of D action. It is also highlighted that the TL tuning leads to a comparatively sluggish but non- oscillatory response due to the more conservative tuning parameters. Large and sudden changes in the control input are not desirable in the process industry to avoid hitting operating constraints (e.g. flooding / weeping in sieve tray towers) during transients. Also, the process equipment changes its dynamic characteristics due to equipment fouling, change in process through-put, wear and tear over time etc so that the need for retuning a control loop is mitigated using conservative controller settings. The TL settings thus represent a good compromise between control performance and robustness.

Process Identification

Obtaining the ultimate gain and period of a control loop by increasing the controller gain causes the process to be driven towards instability. Considering the hazardous nature of chemicals processed in any chemical plant, such a methodology for tuning loops must be avoided. Alternative methods are needed that can be used for proper tuning. Two practical methodologies namely, the process reaction curve and auto-tune variation are presented next.

Process Reaction Curve Fitting

The process reaction curve is the open-loop response of the output variable to a step change in the manipulated variable which usually corresponds to a step change in a valve position. Most of the transient responses can be well represented by a first order plus dead time model. The model parameters are obtained as illustrated in Figure. The model parameters can be obtained by two methods as illustrated in Figure. In both methods, the ratio of the change in the controlled variable (output) from the initial to the final steady state to the magnitude of the step change gives the process gain KP. For the controller, both input and output are 4-20 mA signals corresponding to the sensor and final control element span. In most commercial DCS systems, this range is represented as an equivalent 0-100% range. The units of KP are then % change in controlled variable per % change in manipulated variable.

The two methods differ in the manner in which the dead time, θ, and the first order time constant, τ_p, are obtained. In Method 1, a tangent at the inflection point in the process reaction curve is drawn. Its intersection with the time axis gives the dead time θ. Its intersection with the horizontal line $Y = Y_{ss}$, where Y_{ss} is the final steady state equals $\theta + \tau_p$, from where τ_p is obtained. Equivalently, τ_p is obtained as

$$\tau_P = \frac{K_p}{S}$$

where S is the slope of the tangent drawn at the inflection point.

In Method 2, the time it takes for the response to reach 28.3% and 63.2% of the final steady state are noted. Denote these two times with $t_{28.3\%}$ and $t_{63.2\%}$ respectively. Noting that for a first order lag, 28.3% and 63.2% response completion occurs in $\tau_p/3$ and τ_p time units respectively, we have

$$\theta + \tau_p / 3 = t_{28.3\%}$$

$$\theta + \tau_p = t_{63.2\%}$$

Subtracting the two equations to eliminate θ, we have

$$\tau_p = 1.5\left(t_{63.2\%} - t_{28.3\%}\right)$$

and finally
$$\theta = 1.5 t_{28.3\%} - 0.5 t_{63.2\%}$$

The response of the fitted model using the two methods in shown in Figure. Method 2 is clearly simpler and fits the actual process reaction curve better.

Fitting a First Order Plus Dead Time Model

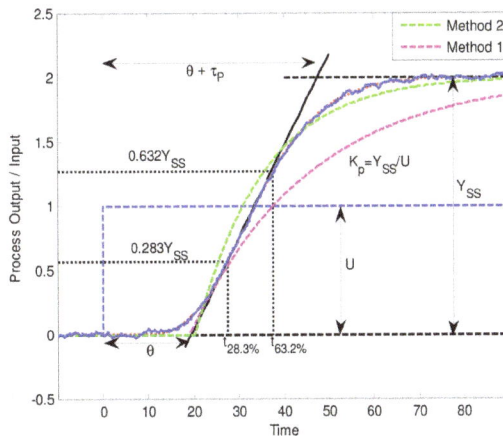

Fitting a first order plus dead time model to the process reaction curve

With the fitted model, K_U and P_U can be obtained either by simulation or complex variable analysis.

Autotuning

Astrom and Hagglund (1984) proposed a powerful auto-tune variation (ATV) method for obtaining the ultimate gain and ultimate period. The method consists of putting a relay at the error signal that toggles the process input by ±h% on detecting a zero crossing. This is schematically illustrated in Figure (a). The action of the relay causes the process input to toggle around the steady state by ±h% for every zero crossing in the error signal corresponding to the output crossing the set-point. Sustained oscillations result and the system ends up in a limit cycle as depicted in Figure (b). The period of oscillations is the ultimate period P_U. The amplitude a of the output oscillations gives the ultimate gain K_U as

$$K_u = \frac{4h}{a\pi}$$

The ATV method has advantages over open loop step methods. The method automatically finds the critical frequency (or period) of the process. Also, large deviations away from the steady state are avoided as this is a closed loop test. Finally, the amplitude at the critical frequency (ultimate period) is obtained so that the identification procedure is more accurate than step / pulse tests.

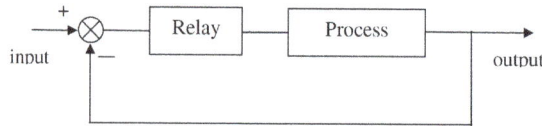

(a). Block diagram of relay feedback approach

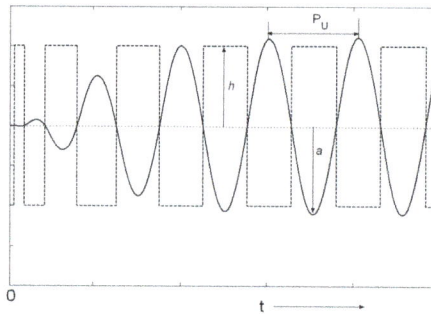

(b). Relay feed back experiment a process with positive steady state gain

Controller Modes and Action

In all DCS systems, the controller can be in the indicator, manual, automatic or cascade mode. In the indicator mode, the controller is off and the process variable (controlled variable) is displayed. The control valve position cannot be adjusted by the operator. In the manual mode, the controller is off. The process variable reading is displayed and the operator can manually input the control valve position. Open loop step / pulse tests are performed in the manual mode with the operator giving a step change to the control valve position. In the automatic mode, the controller is on so that the control valve position is now set by the controller. The operator inputs the set-point for the controlled variable. In the cascade mode, the controller receives the set-point for the controlled variable from a master controller (and not the operator).

Depending on the sign of the process gain, the controller action must be specified to be "direct" or "reverse". Usually a "direct" acting controller increases the controller output as the controlled variable increases above the set-point. A reverse acting controller, on the other hand, decreases the controller output as the controlled variable increases above set-point. For a negative process gain, the controller is "direct" acting while for a positive process gain the controller is "reverse" acting. The definition of "direct" or "reverse" action can vary from one vendor to the other and it is always best to confirm the definition. Another consideration in correctly specifying the controller action is whether the control valve fails open (air-to-close) or fails closed (air-to-open). Process safety considerations dictate if a control valve fails open or fails closed. For example the cooling water valve for removing heat from a reactor would fail open while the steam valve into a reboiler would fail close. If the controller action for a fail open valve is "direct", the action would be "reverse" for a fail close valve in the same control loop.

In control parlance, the controller gain is many-a-times reported as proportional band.

The proportional band is defined as

$$PB = \frac{100}{K_C}\%$$

The higher the proportional band, the lower the controller gain and vice versa.

Rules of Thumb for Controller Tuning

Almost all control loops in the process industry are one of the following

> Flow control loop
>
> Pressure control loop
>
> Level control loop
>
> Temperature control loop
>
> Product Quality Control Loop

Some heuristics are discussed for tuning these loops that reflect common industrial practice. Depending on the application, exceptions to these heuristics are always possible.

Flow Loops

Flow is usually controlled using a PI controller. The signal from the flow sensor is noisy due to turbulent flow so that a large proportional band (about 150%) is used. A small reset time (10-20s) is used for good set-point tracking.

Level Loops

Most liquid levels provide surge capacity for filtering out flow disturbances. For example, the reflux drum in a distillation column allows for the reflux into the column to be held constant even as the vapour condensation rate and distillate rate vary. If the drum is not provided, the reflux into the column would fluctuate unnecessarily disturbing the column. The reflux drum thus acts as a surge capacity. In order to filter out flow disturbances, the level should be controlled loosely. The control objective is to maintain the liquid level within acceptable limits. Accordingly, a P controller is used for level control. A proportional band of 50% is commonly used so that the valve fully closes / opens for a 25% change in the level assuming the valve is initially 50% open. Note the use of PI controllers for level control of surge capacities is not recommended as a change in the inlet (outlet) flow would require that the outlet (inlet) flow increase above (decrease below) the inlet flow in order to bring the level back to its set-point (zero offset). The flow disturbance thus gets magnified downstream (upstream). This magnification would only worsen for a series of interconnected units defeating the very purpose of providing surge capacity for attenuating flow disturbances. There are, of course, exceptions where tight level control is desired. For example, the level in a CSTR should be controlled tightly to maintain the residence time.

Pressure Loops

The dynamics of pressure in a can be very fast (flow like) or slow (level like) depending on the process system. For example, the pressure dynamics are extremely fast for a valve throttling the vapour outlet line from a tank. On the other hand, the dynamics are slow for the cooling water flow adjusting the pressure in a condenser due to the heat transfer and water flow lag. PI controllers are usually used for pressure loops with a small proportional band (10-20%) and integral time (0.2-2 mins) for tight pressure control. Tight pressure control is usually desired in most processing situations. For example, in distillation columns, the pressure must be controlled tightly as large pressure deviations would require compensation of the temperature controller set-points that ensure inferential product quality control. Similarly, most gas phase reactors are designed for near maximum pressure operation for maximum reaction rates so that large pressure deviations are not acceptable.

Temperature Loops

Temperature loops are moderately slow due to sensor lags and heat transfer lags. PI and PID controllers are often used. In most processing situations, tight temperature control is desired so that the proportional band is low (2-20%). The integral time is usually set to about the same value as the process time constant. In situations where derivative action is used for faster closed loop response, the derivative time constant is set to about one-fourth the process time constant or less depending on the transmitter signal noise.

Quality Loops

Composition control loops are usually applied for maintaining the product quality. In terms of relative importance, these loops are probably the most crucial for process profitability. If the product quality shows large variability, the process must be operated at a mean product quality that is significantly better than the quality specification to ensure the production of on-spec or better quality product all the time. This results in a quality giveaway

adversely affecting the process profitability. The quality giveaway can be reduced by ensuring tight product quality control. The concept of quality give-away is illustrated in Figure.

Typical composition measurements involve large dead-times or lags. For example the dead-time introduced by a gas-chromatograph can vary from a few minutes to an hour. Some compositions may be measured once a shift or once a day through laborious analytical measurements. Of all the measurements, analytical composition measurements are the most expensive and unreliable. The product specifications increasingly require the measurement of ppm / ppb levels of trace impurities so that a logarithmic scale is more appropriate in many situations. Product quality measurements are typically used to make small / incremental adjustments in the set-point of a loop. The frequency of the changes may vary from once a day to once every hour etc. Whenever PID controllers are applicable, a large proportional band is used (100-2000%). A large reset time (0.1 – 2 hrs) must be used due to the lag introduced by the composition measurement as well as the usually slow process dynamics.

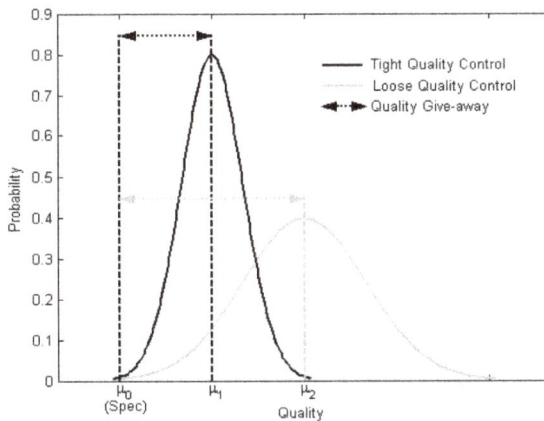

The concept of quality give-away

Advanced Control Structures

The feedback control loop, discussed at length, forms the backbone of control systems applied in the process industry. Some typical feedback control loops are schematically illustrated in Figure. Over the years, enhancements to the basic feedback control structure that lead to significant improvement in control performance, have been developed. These advanced control structures include ratio control, cascade control, feed-forward control, over- ride control and valve positioning control and are briefly described in the following.

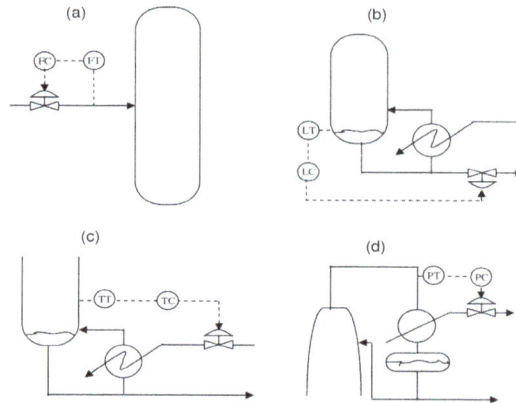

Typical feed back control schemes commonly employed in distillation columns. (a) Feed flow control, (b) Level control in reboiler drum using bottoms flow, (c) Tray temperature control using reboiler duty and (d) Column pressure control using condenser duty.

Ratio Control

Ratio control, as the name suggests, is used for maintaining the ratio between two streams. The independent stream is referred to as the wild stream. The ratio controller adjusts the flow of the other stream to keep it in ratio to the wild stream. The implementation of ratio control is illustrated in Figure. The wild stream flow measurement is multiplied by the ratio set-point to obtain the flow set-point for the manipulated stream. The calculated flow set-point is input to the flow controller on the manipulated stream. Ratio control is implemented as a feed-forward strategy (to be discussed later) where two flows are increased in tandem so that the change in the wild stream is compensated for before it affects the process output. For example, if the feed flow rate into a distillation column increases by 10%, the reboiler duty necessary to maintain the same separation should also increase by about 10%. It therefore makes sense to ratio the reboiler duty to the fresh feed rate so that the necessary change in the reboiler duty is implemented apriori. This leads to tighter product purity control with the change in the feed rate causing only small deviations in the product purity.

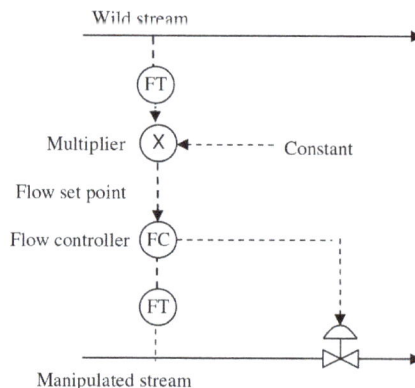

Implementation of ratio control.

Cascade Control

Cascade control is arguably one of the most useful concepts in chemical process control. The cascade control scheme consists of two control loops, namely the master loop and the slave loop, with the master loop setting the set-point for the slave loop. The concept is best illustrated by an example. Consider a jacketed CSTR where cooling water is recirculated in the jacket to remove the exothermic reaction heat. The typical feedback reactor temperature control scheme and the cascade reactor temperature control scheme is shown in Figure. In the feedback arrangement, the reactor temperature controller directly adjusts the cooling water valve to maintain the reactor temperature at set-point. In the cascade arrangement, a slave loop is introduced that controls the jacket temperature by manipulating the cooling water valve. The master reactor temperature loop adjusts the jacket temperature set-point.

At first glance, the advantage of cascade arrangement over simple feedback control is not very obvious. To appreciate the same, consider an increase in the coolant temperature as an input disturbance. In the simple feedback scheme, the reactor temperature must rise before the controller opens the cooling water valve to bring the reactor temperature back to set- point. In the cascade control scheme, the jacket temperature controller senses the increase in the cooling water temperature and adjusts the cooling water valve to maintain the jacket temperature. The reactor temperature would thus show comparatively much smaller / negligible deviations from set-point. The slave controller acts to remove local disturbances into the process and prevents its effect on the primary controlled variable. Another subtle advantage is that the slave controller compensates for the non-linearity in the slave loop so that the master controller 'sees' a more linear system. In the current example, the non-linear characteristics of the cooling water valve are compensated for by the slave controller. Since the slave loop has much faster dynamics than the master loop (else the cascade arrangement is infeasible), the master loop does not have to compensate for the valve non-linearity. It therefore sees a less non-linear system compared to simple feedback control resulting in improved control performance. The improvement is however at the expense of installing, tuning and maintaining an additional slave controller.

To tune a cascade control structure, the slave loop is first tuned with the master loop in manual. P only controllers with a small proportional band (large controller gain) are commonly used in the slave loop for a fast response to a set-point change from the master controller. Integral action is usually not applied in the slave loop as an offset in the secondary measurement is acceptable. The tuned slave loop is then put on automatic and the master loop is tuned. Note that for the cascade control system to be stable, the dynamics of the slave loop should be much faster than the master loop allowing the slave loop to keep-up with the set- point changes received from the master loop. A typical rule of thumb is that the time constant for the master loop should be more than thrice that of the slave loop.

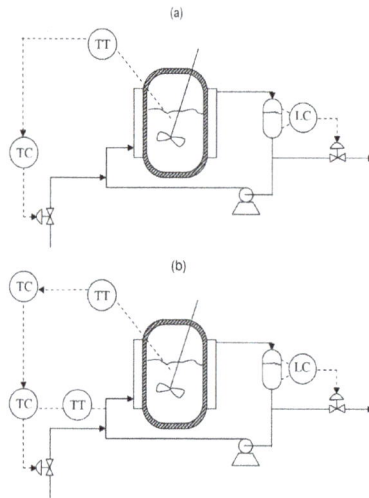

Temperature control of an exothermic CSTR. (a) the typical feedback reactor temperature control scheme and (b) the cascade reactor temperature control scheme.

Cascade control loops are quite common in the process industry. Some common configurations are shown in Figure. The interpretation of these configurations is left as an exercise to the reader.

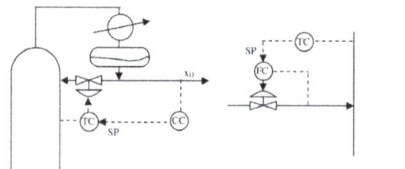

Some typical cascade arrangements

Feed-forward Control

The concept of feed-forward control has already been alluded to earlier. If a measured disturbance enters a process, the control input can be adjusted to compensate for effect of the disturbance on the output. Perfect compensation would cause the controlled output to show no deviations from its set-point even as a disturbance has entered the process. This apriori compensation to mitigate the transient effect of a measured disturbance on the controlled output is referred to as feed-forward control. A very simple example of feed-forward control is driving a car. Adjusting the hot and cold water knobs for the right temperature water from the shower is an example of feedback control. As discussed previously, ratio control compensates for disturbances in a feed-forward manner.

The design of a feed-forward compensator is illustrated using block diagrams in Figure. G_d represents the disturbance to output transfer function while G_p represents the control input to output transfer function. The control input u must be varied such that

$$G_p.u + G_d.d = 0$$

The control input is adjusted by the feed-forward compensator with the transfer function G_{ff} so that

$$u = G_{ff}.d.$$

Substituting into the previous equation and solving for G_{ff} gives the feed-forward compensator design as

$$G_{ff} = -G_d / G_p$$

Assuming that G_d and G_p are first order plus dead time transfer function, the feed-forward compensator is then a lead-lag plus dead time transfer function. Modern DCS allow lead-lag plus dead time blocks to be configured into the control system.

For a better appreciation of the improvement in control performance using feed-forward compensation, consider a very simple example where

and
$$G_d = 1/(s+1)$$
$$G_p = 1/(5s+1)$$

Then
$$G_{ff} = -(5s+1)/(s+1)$$

Figure plots the simulated transient output response for a unit step change in the measured disturbance with and without feed-forward compensation.

Since there is no plant-model mismatch, perfect feed-forward compensation is observed with the output showing no deviations from set-point. In a real-life scenario, the presence of a plant-model mismatch may cause small transient deviations. The feedback controller compensates for these small deviations resulting in an overall tighter closed loop response.

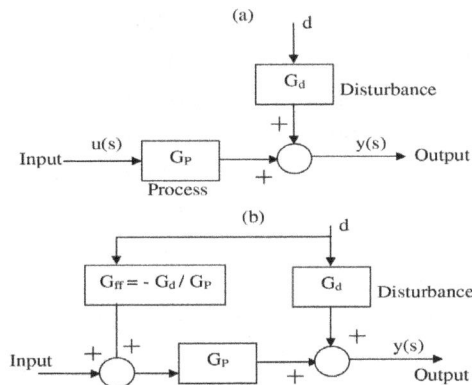

Design of feed forward compensator. (a) Process and
(b) process with feed forward compensator.

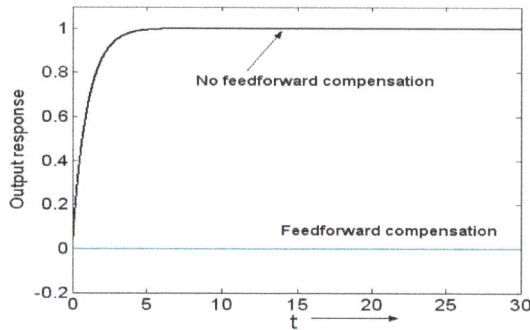

Deviation in the output with and without feed forward action

Override Control

Over-ride control is employed to ensure that an unsafe condition does not arise during process operation. As the name suggests, an over-ride controller over-rides the output of another controller as an unsafe condition develops and acts to move the process away from the unsafe condition. This is an example of multivariable control where the same manipulated variable can be adjusted at any time by one of many controlled variables. An example best illustrates the concept of over-ride or selective control. Consider the bottom section of a distillation column. The bottom sump level is controlled by the bottoms flow rate. During normal operation, the steam rate into the reboiler is manipulated to control a tray temperature. During severe transients, a situation may arise where the bottoms level is low and continues to fall even as the bottoms flow rate is zero. An unsafe situation can arise with the reboiler tubes getting exposed to vapour and fouling. Also, the bottoms pump may lose suction as the reboiler dries up. A sensible operator would put the temperature loop on manual and cut back on the steam rate to ensure the reboiler tubes remain submerged. In effect, the temperature controller output, the signal to the steam valve, gets over-ridden to maintain the liquid level. The over-ride controller automates this action as shown in Figure. The base level signal is input to a multiplier. A multiplier value of 5 is used so that if the level is above 20%, the multiplier output is above 100%. As the level decreases below 20%, the multiplier output decreases below 100%. If the level continues to decrease, the multiplier output would eventually decrease below the temperature controller output. The low select would then pass on the multiplier signal to the steam valve over-riding the temperature controller. The steam rate would thus decrease. Once the level begins to rise, the multiplier output would increase above the temperature controller output so that the low select would pass the manipulation of the steam valve back to the temperature controller. In addition to the level over-ride controller, the low select may also receive signals from a pressure over-ride controller or a temperature over-ride controller to reduce the steam flow rate. Pressure over-ride would be needed if the column pressure goes too high. Similarly temperature over-ride may be necessary if the base temperature goes too high.

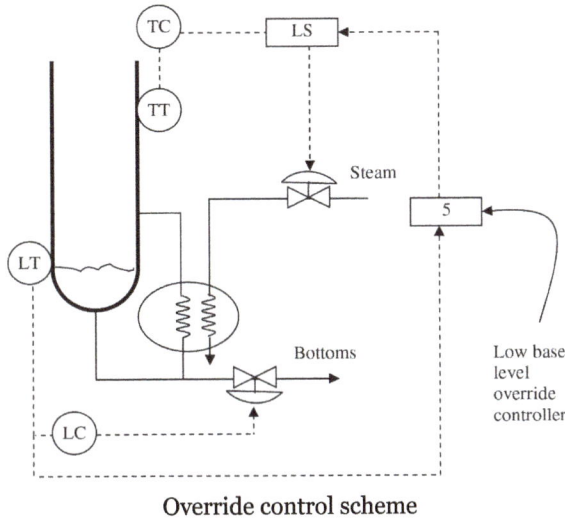

Override control scheme

In temperature or pressure over-rides, a PI controller is needed unlike the P only controller for a level over-ride. This is because a pressure / temperature over-ride is needed only for a very small range of the total transmitter span. A very large proportional gain would then be necessary which can destabilize the closed loop system. Therefore a PI controller with lower gain and fast reset action is used to achieve the tightest control possible.

Valve Positioning (Optimizing) Control

Valve positioning control was originally proposed by Shinskey as an effective way of minimizing the energy consumption in distillation columns. The pressure in a distillation column is set by the condenser cooling duty. For a given separation, as the column pressure increases, more stages are needed as the x-y VLE plot moves towards the 45 degree line as shown in Figure. Translated to process operation, the same separation can be achieved at lower reboil as the column operating pressure is reduced. To minimize energy consumption, the column should be operated at lowest possible pressure corresponding to the maximum condenser duty. This can be accomplished by the valve positioning control scheme as illustrated Figure. The column pressure is typically controlled by adjusting the condenser cooling water valve. The VPC controller takes in the pressure controller output signal and adjusts the pressure set-point. If the valve is not nearly open, the controller reduces the column pressure set-point so that the pressure controller increases the cooling duty to reduce the column pressure. The VPC controller thus ensures that any underutilized cooling capacity is exploited to reduce the column operating pressure. The column pressure thus floats with the condenser duty being near maximum. The VPC controller is tuned to be slow with the fast pressure controller rejecting any pressure disturbances.

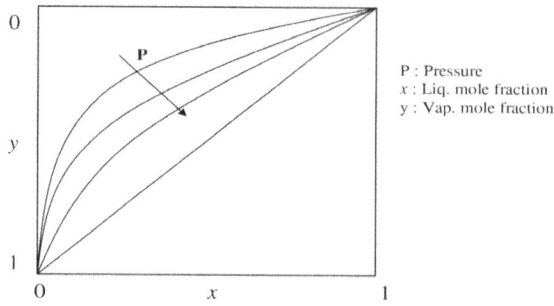

Typical x-y diagram with varying Pressures

P : Pressure
x : Liq. mole fraction
y : Vap. mole fraction

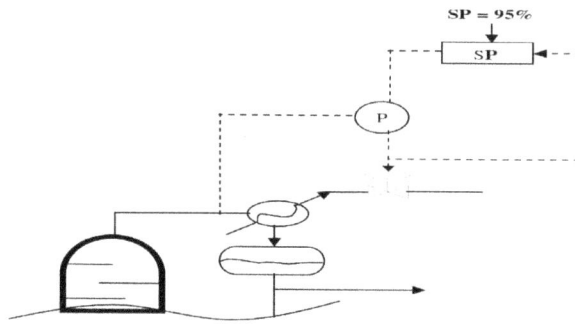

Valve positioning control

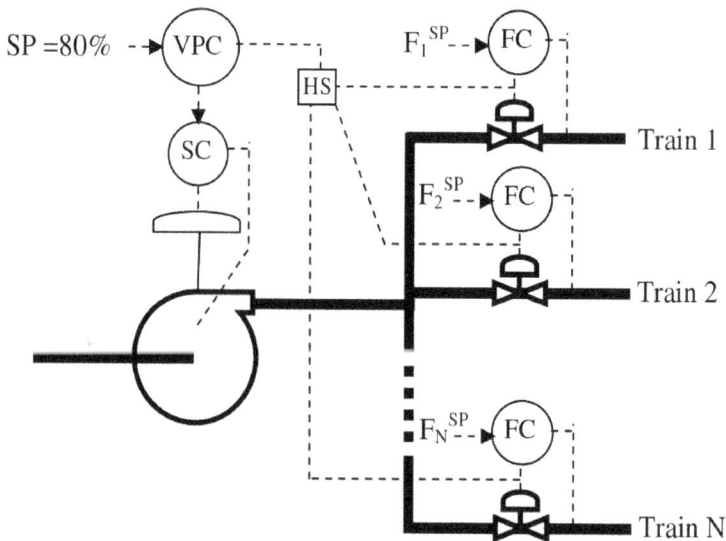

VPC for minimizing variable speed pump electricity

Another simple VPC application is shown in Figure. Let us say a high capacity variable speed pump is providing feed to N parallel trains of processes. We would like to minimize the pump electricity consumption while ensuring the desired flow setpoints for each of the parallel trains is achieved. The electricity consumption gets minimized by running the pump at as low an rpm as possible. This gets achieved by ensuring that the most open process feed valve is nearly fully open. The high select passes the position of the most open valve. A valve position below the nearly fully open VPC setpoint (say 80%) indicates unnecessary valve throttling. The VPC then reduces the pump rpm. In response, the flow controllers would open the valves to maintain the flow. The VPC reduces the pump rpm till the most open valve position reaches the VPC setpoint (80%) ensuring the pump operates at as low an rpm as possible while maintaining the desired flow to each of the parallel trains.

Multivariable Systems

Single input single output (SISO) systems have been treated till now. Most practical control system design problems are multivariable in nature with multiple inputs multiple outputs (MIMO). A 2 X 2 multivariable system is shown in Figure. There are two inputs, u_1 and u_2 and two outputs y_1 and y_2. In the most general case, a step change in an input causes a transient response in both the outputs. The input output relationship may be compactly represented in matrix notation as

$$\begin{bmatrix} y_1(s) \\ y_2(s) \end{bmatrix} = \begin{bmatrix} G_{11}(s) & G_{12}(s) \\ G_{21}(s) & G_{22}(s) \end{bmatrix} \begin{bmatrix} u_1(s) \\ u_2(s) \end{bmatrix}$$

and the corresponding block diagram is shown in Figure.

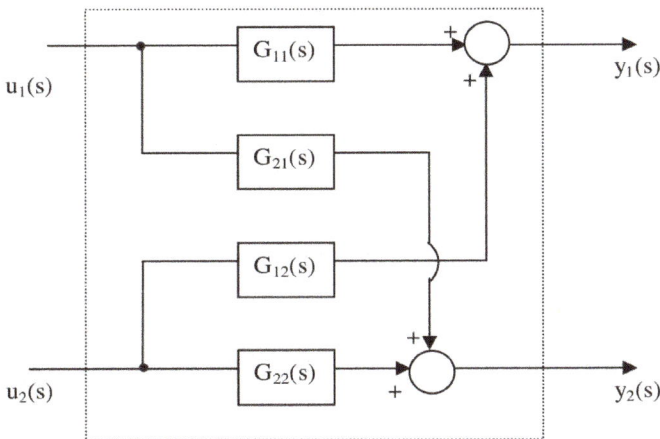

A block diagram of a 2 X 2 multi variable system

In general, G_{ij} denotes the transfer function between the jth input and the ith output. The non-diagonal terms with $i \neq j$ are the interaction terms. The simplest way of controlling a multivariable process is to control each of the outputs by manipulating an input using a PID controller. This is referred to as multivariable decentralized control and is illustrated in Figure. for the example 2x2 system. Controller 1 manipulates u_1 to maintain y_1 and controller 2 adjusts u_2 to maintain y_2.

In the design of a multivariable decentralized control system, choice exists as to which manipulated variable is used to control an output. For the 2x2 example, there are a total of two control structures with y_1 being controlled by u_1 or u_2. The number of such possibilities grows exponentially as the number of inputs / outputs increase. In the most general sense, the design of a plant-wide decentralized control system for a complex chemical process is a multivariable problem of high order. The high order problem is naturally broken down into smaller process unit specific controller design problems and controller design for managing plant-wide issues such as inventory balancing. A high order unit specific controller design problem can also be further broken down into a smaller subset of fast loops and slow loops based on the process dynamics. An example is the simplification of the 5x5 controller design problem for a simple distillation column into a 2x2 problem. In a distillation column, the pressure, reflux drum and bottom levels and two temperatures (or compositions) may be controlled. Since the tray temperature dynamics are significantly slower than the pressure / level dynamics, SISO controllers are applied for the latter reducing the 5x5 problem into a 2x2 design problem for the two temperature controllers. Any complex high order control system design problem can thus be simplified into subsets of simple SISO, 2x2 or in the worst case 3x3 decentralized control system design problems. A systematic unit specific and plant-wide control system design methodology for complete chemical plants will be developed in the subsequent chapters.

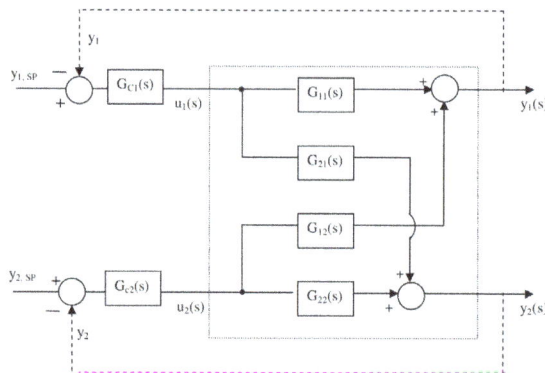

Block diagram of a multivariable decentralized control for a 2X2 system

Interaction Metrics

The selection of the input-output pairing in a decentralized control system is usually made based on engineering considerations which shall be covered in greater detail in subsequent chapters. The individual controllers in a decentralized control system may need to be detuned in order to maintain process stability. This is because the interaction between the loops during closed loop operation can lead to instability. The magnitude of interaction depends on the aggressiveness of the individual controller tunings employed. Detuning or less aggressive tuning mitigates the interaction to ensure closed loop stability. The Niederlinski Index and Relative Gain Array are two commonly used quantitative measures of interaction between control loops. Both are based on the open-loop steady state gain matrix K_p, where

$$y = \mathbf{K_p u}$$

Niederlinski Index

The Niederlinski Index for a control structure where the ith input is used to control the ith output is then defined as

$$NI = \frac{\left|K_p\right|}{\prod_i K_{ii}}$$

The NI for any control structure can thus be obtained through appropriate relabeling of the outputs and inputs so that the ith input controls the ith output. If the Niederlinski Index is negative, the closed loop system is guaranteed to be integral closed loop unstable. If the NI is positive, the closed loop system may or may not be stable. In other words, the criteria NI>0 is a necessary but not sufficient condition for closed loop stability. Input-output pairings with small positive or large positive (>>1) NI values indicate ill-conditioning problems and should be avoided. Control structures with NI close to 1 indicate favourable interaction. For example, an NI value of 1 for a 2X2 system indicates that either K_{12} or K_{21} or both are zero implying one-way or no steady state interaction between the loops. The primary use Niederlinski Index is for rejecting unworkable control structures.

Relative Gain Array

The relative gain is another popular metric that measures the interaction of a control loop with other loops as the ratio of the steady state process gain the controller sees with all other loops off to the process gain with all other loops on (all other outputs at their set- points). Mathematically, if the ith output is controlled by the jth input, its relative gain is defined as

$$\lambda_{ij} = \frac{\left(\dfrac{\partial y_i}{\partial u_j}\right)_{u_k=constant, k \neq j}}{\left(\dfrac{\partial y_i}{\partial u_j}\right)_{y_k=constant, k \neq i}}$$

If the relative gain is negative, the ith output should not be paired with the jth input as the process gain sign would change depending on whether the other loops are on automatic or manual mode. Input-output pairings with relative gain close to 1 may be preferred as the process gain the controller sees is independent of the state of the other loops. The relative gain array is obtained as i and j are varied for respectively all outputs and inputs.

The relative gain array is an effective tool for input-output pairing when the primary control objective is set-point tracking. For set-point tracking, lower interaction between the loops increases the degree of independence of the different control loops so that each can be separately tuned for tight set-point tracking. Interaction is thus undesirable for set-point tracking. For load disturbance rejection, interaction is not necessarily undesirable and may actually favour disturbance rejection. This was demonstrated in an early article by Niederlinski (1971). Since the primary objective in chemical process control is load rejection, the application of RGA for control structure selection makes little sense. Candidate control structures should be proposed based on engineering considerations and unworkable structures further eliminated using the Niederlinski Index. The same arguments can be applied to recommend the use of dynamic decouplers only when the primary control objective is set- point tracking. Dynamic decoupling is not covered here as load rejection is the primary control objective in chemical process control systems.

Multivariable Decentralized Control

Consider the 2x2 multivariable open loop system in Figure. We would like to hold both the outputs at their respective setpoints. The simplest way to do it is to implement individual PI controllers for y_1 and y_2. Without loss of generality, let us assume that y_1 is paired with u_1 and y_2 is paired with u_2. The multivariable control system is shown in Figure. Notice that even as u_1 and u_2 affect both y_1 and y_2 through the interaction transfer functions G_{12} and G_{21}, the adjustment made to u_1 is based purely on e_1 and the adjustment made to u_2 is based purely on e_2. In other words, the y_1 controller moves are based purely on y_1 and does not consider the effect of the control moves made by the y_2 controller. Similarly, the y_2 controller moves are based purely on y_2 and does not consider the effect of control moves made by the y_1 controller. Thus even as the actual system is multivariable, the individual controllers do not take the interaction into consideration. This is referred to as decentralized control.

For the decentralized control system, notice that the interaction terms introduce an additional feedback path as shown in blue in Figure. This additional feedback tends to further destabilize the closed multivariable control system. If each controller is tuned individually with the other controller on manual (other loop is open) and the Zeigler Nichols tunings applied, then when both the loops are closed, the system response is likely to be highly oscillatory and may even be unstable due to the additional feedback path. In the individual tuning of the controllers, since the other loop is open, this additional feedback path is inactive and therefore not accounted for in the determination of the tuning parameters. Clearly the individual ZN tuning parameters need to be detuned due to the additional feedback path to ensure the overall closed loop response is sufficiently away from instability.

Detuning Multivariable Decentralized Controllers

The obvious next question is that how does one tune a decentralized multivariable controller. Typically, in practical settings, tight control of one of the outputs is much more important than the other. A sequential tuning procedure can then be applied, where the more important output controller is tuned individually so that we get the tightest possible controller tuning. The less important output controller is then tuned with the other loop on automatic. Since the other loop is on, the additional feedback path is active and the necessary detuning due to the same gets accounted for in the tuning parameters of this less important loop. This sequential tuning procedure thus gives the tightest possible control of the more important output at the expense of a highly detuned controller for the less important output. The sequential procedure can be easily extended to more than 2 outputs when the prioritization of the controlled outputs is clear.

There are however situations where the need for tight control of each of the outputs is comparable. The detuning due to multivariable interaction then needs to be taken in all the loops. How does one systematically go about the detuning. For the 2x2 multivariable system, we have for the open loop system

$$\begin{bmatrix} y_1 \\ y_2 \end{bmatrix} = \begin{bmatrix} G_{11} & G_{12} \\ G_{21} & G_{22} \end{bmatrix} \begin{bmatrix} u_1 \\ u_2 \end{bmatrix}$$

or more simply

$$y = G_p u$$

where G_p is the open loop process transfer function matrix. For a decentralized controller, we have

$$\begin{bmatrix} u_1 \\ u_2 \end{bmatrix} = \begin{bmatrix} G_{C1} & 0 \\ 0 & G_{C2} \end{bmatrix} \begin{bmatrix} y_1^{sp} - y_1 \\ y_2^{sp} - y_2 \end{bmatrix}$$

or in matrix notation

$$u = G_C (y^{SP} - y)$$

where the controller matrix, G_C, is diagonal for decentralized control. Combining the above two matrix equations, we get

$$y = G_p G_C (y^{SP} - y)$$

or
$$(\mathbf{I}+\mathbf{G_P}\,\mathbf{G_C})\mathbf{y} = \mathbf{G_P}\mathbf{G_C}\mathbf{y}^{SP}$$

or
$$y = (\mathbf{I}+\mathbf{G_P}\mathbf{G_C})^{-1}\,\mathbf{G_P}\mathbf{G_C}\mathbf{y}^{SP}$$

This is the multivariable closed loop servo response equation and its analogy with SISO systems is self evident. Each element of the $(I+\mathbf{G_P}\mathbf{G_C})^{-1}$ matrix would have det $(I+G_pG_C)$ as its denominator. The closed loop multivariable characteristic equation is then

$$det(I + G_pG_C) = 0$$

Similar to SISO systems, if any of the roots of the multivariable characteristic equation is in the right half plane, the closed loop multivariable system is unstable.

To systematically detune the controllers, an empirical analogy with the Nyquist stability criterion for SISO systems is used. For a SISO system, the closed loop servo response equation is

$$y = [G_pG_C / (1+G_p\,G_C)]y^{SP}$$

where G_P is the open loop transfer function and G_C is the controller transfer function. The Nyquist stability criterion then guarantees stability for the closed loops system if the polar plot of the open loop transfer function between y^{SP} and y, ie G_pG_C, does not encircle (-1, 0). Gain margin and phase margin are criteria that are commonly used to quantify the distance from (-1, 0) at a particular frequency. To ensure that the distance from (-1, 0) is sufficient at

all frequencies, the 2 dB closed loop maximum log modulus criterion is often used, where the closed loop log modulus is defined as

$$L_{CL}(\omega) \;=\; 20\,log\big|G_pG_C \,/\, (1+G_pG_C)\big|_{s=j\omega}$$

L_{CL} is calculated by putting $s = j\omega$ in the transfer functions, G_p and G_C, and is therefore a function of ω. The SISO PI tuning parameters (K_C and ι_i) are chosen such that the maximum closed loop log modulus (with respect to ω) is 2dB. This ensures that the closed loop servo response is fast and not-too-oscillatory.

To develop a closed loop maximum log modulus criterion for multivariable systems, we note that the SISO closed loop characteristic equation is

$$1 + G_pG_C = 0$$

and the transfer function whose polar plot is used to see encirclements of (-1,0) is then

$$-1 + (1 + G_pG_C)$$

ie $-1+ closed\ loop\ characteristic\ equation$

For a multivariable system, we then define by analogy

$$W = -1 + det(I + G_P G_C)$$

where W is $-1+ closed\ loop\ characteristic\ equation$. The multivariable closed loop log modulus (L_{MVCL}) is then defined as

$$L_{MVCL} = 20log|W/(1+W)|.$$

The tuning parameters for the individual controllers should be chosen such that

$$L_{MVCL}{}^{MAX} = 2 N_C$$

where N_C is the number of loops.

A simple algorithm for systematic detuning of the individual controller for the 2x2 decentralized control system is then:

1. Obtain individual ZN tuning parameters, $K_{C1}{}^{ZN}, \tau_{I1}{}^{ZN})$ and $(K_{C2}{}^{ZN}, \tau_{I2}{}^{ZN})$, for each loop.

2. Detune the individual tuning parameters by a factor f $(f > 1)$ to get the revised tuning parameters as $\left(K_{C1}{}^{ZN}/f, f.\tau_{I1}{}^{ZN}\right)$ and $\left(K_{C2}{}^{ZN}/f, f.\tau_{I2}{}^{ZN}\right)$

3. Adjust f such that $L_{MVCL}{}^{MAX} = 4\ dB$.

The above procedure can be easily extended to an NxN (N > 2) decentralized control system.

As a parting thought, we re-emphasize that in chemical processes, the dominant time constants of different loops can differ by up to two orders of magnitudes. Thus for example, the residence time of a surge drum may be ~5 minutes while it may take 2-5 hrs for transients caused by a change in its setpoint to reach back after passing through the different downstream units, the material recycle and the upstream units. Similarly, on a distillation column, while the column pressure time constant with respect to condenser duty is ~1 min and the reflux drum / bottom sump level residence times are ~ 5 mins, the tray temperature response times to changes in reflux / boilup rates are much slower (~15-20 mins). Thus even as the dual-ended distillation column control problem is 5x5 (2 levels, 1 pressure and 2 temperatures), the separation in time constants allows the level and pressure controllers to be tuned first followed by the two temperature controllers. The 5x5 problem thus reduces to a 2x2 problem due to the separation in time constants. In industrial practice, most high order multivariable problems reduce to 2x2 or at most 3x3 problems, which are mathematically tractable.

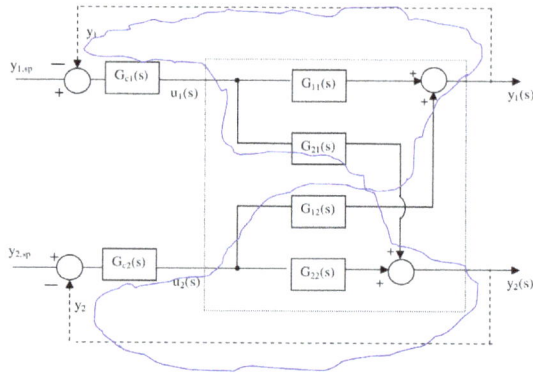

Additional feedback path due to multivariable interaction

Illustrative Example:

Consider a 2x2 openloop multivariable system

$$\begin{bmatrix} y_1 \\ y_2 \end{bmatrix} = \begin{bmatrix} \dfrac{-18.9e^{-3s}}{21s+1} & \dfrac{12.8e^{-s}}{16.7s+1} \\ \dfrac{-19.4e^{-3s}}{14.4s+1} & \dfrac{6.6e^{-7s}}{10.9s+1} \end{bmatrix} \begin{bmatrix} u_1 \\ u_2 \end{bmatrix}$$

(a) Calculate its RGA. Based on the RGA, what input-output pairing would you recommend.

(b) Calculate the Niederlinski Index for the recommended pairing. What can you say about closed loop integral stability of the recommended pairing.

(c) Calculate the Niederlinki Index for the other alternative pairing (the one that is not recommended). What can you say about the closed loop integral stability of this other pairing.

(d) For the recommended pairing, design a feedforward dynamic decoupler showing its complete block diagram and also the physically realizable feedforward compensator transfer functions.

Solution:

(a) The steady state input-output relationship is

$$\begin{bmatrix} y_1 \\ y_2 \end{bmatrix} = \begin{bmatrix} -18.9 & 12.8 \\ -19.4 & 6.6 \end{bmatrix} \begin{bmatrix} u_1 \\ u_2 \end{bmatrix}$$

so that the steady state gain matrix is

$$K = \begin{bmatrix} -18.9 & 12.8 \\ -19.4 & 6.6 \end{bmatrix}$$

Inverting the matrix, we get

$$K^{-1} = \begin{bmatrix} 0.0534 & -0.1036 \\ 0.1570 & -0.1529 \end{bmatrix}$$

The RGA is then obtained as

$$RGA = \mathbf{K}.*(\mathbf{K}^{-1})^T$$

where the '.*' operator denotes element-by-element multiplication. Performing the necessary operations, we get

$$RGA = \begin{bmatrix} -1.0094 & 2.0094 \\ 2.0094 & -1.0094 \end{bmatrix}$$

Notice that the row/column sum of the RGA is 1. This is a property of the RGA.

Rejecting the IO pairings corresponding to the negative RGA elements, the recommended pairing based on the RGA is y_1-u_2 and y_2-u_1.

(b) The steady state IO relation for the recommended pairing

$$\begin{bmatrix} y_1 \\ y_2 \end{bmatrix} = \begin{bmatrix} 12.8 & -18.9 \\ 6.6 & -19.4 \end{bmatrix} \begin{bmatrix} u_2 \\ u_1 \end{bmatrix}$$

is The Niederlinski Index is then

$$NI = \frac{12.8 \times (-19.4) - 6.6 \times (-18.9)}{12.8 \times (-19.4)} = 0.4977$$

Since NI > 0 for the recommended pairing, the multivariable decentralized control system may be integrally stable.

(c) The other possible pairing is y_1-u_1 and y_2-u_2. For this pairing, the IO relation is

$$\begin{bmatrix} y_1 \\ y_2 \end{bmatrix} = \begin{bmatrix} -18.9 & 12.8 \\ -19.4 & 6.6 \end{bmatrix} \begin{bmatrix} u_1 \\ u_2 \end{bmatrix}$$

The Niederlinski Index is then

$$NI = \frac{6.6 \times (-18.9) - 12.8 \times (-19.4)}{(-18.9) \times 6.6} = -0.99$$

Since the NI for this pairing is < o, the multivariable decentralized control system is guaranteed to be integrally unstable. This pairing should therefore not be implemented.

(d) If we look at the open loop 2x2 system with the recommended pairing (y_1-u_2 and y_2-u_1), a change in u_2 affects both y_1 (its controlled variable, CV) and y_2 (other CV). Similarly, a change in u_1 affects both y_2 (its CV) and y_1 (other CV). When both the control loops are on, the adjustment made by a loop ends up disturbing the other loop. A dynamic decoupler uses feedforward compensation ideas to make appropriate adjustments in the "other" process input so that a change in a process input only affects its CV and not the other CV. The dynamic decoupler block diagram for the recommended pairing is shown in Figure. We are looking for the feedforward compensator G_I^{ff} $\left(G_{II}^{ff}\right)$ so that a change in u_2 (u_1) only affects its CV, y_1 (y_2) with no effect on the other CV y_2 (y_1).

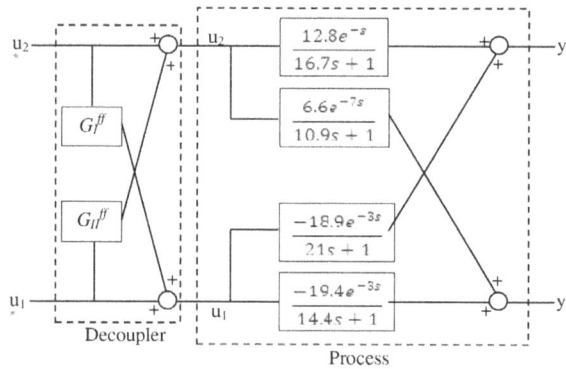

2x2 process example with dynamic decoupler

From the block diagram, the ideal compensator G_I^{ff} would be such that

$$y_2 = G_{22}u_2{}^* + G_{21}G_I^{ff}\,u_2{}^* = 0$$

so that
$$G_I^{ff} = -G_{22}/G_{21}$$

Similarly, we have
$$G_{II}^{ff} = -G_{11}/G_{12}$$

Putting in the appropriate transfer functions, we get

$$G_I^{ff} = -\frac{\dfrac{6.6e^{-7s}}{10.9s+1}}{\dfrac{-19.4e^{-3s}}{14.4s+1}} = 0.3402\,\frac{14.4s+1}{10.9s+1}e^{-4s}$$

$$G_{II}^{ff} = -\frac{\dfrac{-18.9e^{-3s}}{21s+1}}{\dfrac{12.8e^{-s}}{16.7s+1}} = 1.4766\frac{16.7s+1}{21s+1}e^{-2s}$$

The feedforward compensators consist of a gain, a lead-lag and a deadtime. In some cases, it is possible that we get an exponential term of form e^{+Ds} (D > 0) implying a negative dead- time. This means that a change in the causal variable leads to a change in the effected variable in the past, which is impossible. The term e^{+Ds} is then physically unrealizable and dropped from the compensator.

Unit Operations in Chemical Plant

Chemicals can be manufactured on a large scale in chemical plants. They use specific equipment and technology in manufacturing these chemicals. The chapter on unit operations in a chemical plant offers an insightful focus, keeping in mind the complex subject matter.

Chemical Plant

A chemical plant is an industrial process plant that manufactures (or otherwise processes) chemicals, usually on a large scale. The general objective of a chemical plant is to create new material wealth via the chemical or biological transformation and or separation of materials. Chemical plants use specialized equipment, units, and technology in the manufacturing process. Other kinds of plants, such as polymer, pharmaceutical, food, and some beverage production facilities, power plants, oil refineries or other refineries, natural gas processing and biochemical plants, water and wastewater treatment, and pollution control equipment use many technologies that have similarities to chemical plant technology such as fluid systems and chemical reactor systems. Some would consider an oil refinery or a pharmaceutical or polymer manufacturer to be effectively a chemical plant.

BASF Chemical Plant Portsmouth Site in the West Norfolk area of Portsmouth, Virginia, United States. The plant is served by the Commonwealth Railway.

Petrochemical plants (plants using chemicals from petroleum as a raw material or *feedstock*) are usually located adjacent to an oil refinery to minimize transportation costs for the feedstocks produced by the refinery. Speciality chemical and fine chemical plants are usually much smaller and not as sensitive to location. Tools have been developed for converting a base project cost from one geographic location to another.

Chemical Processes

Chemical plants use chemical processes, which are detailed industrial-scale methods, to transform feedstock chemicals into products. The same chemical process can be used at more than one chemical plant, with possibly differently scaled capacities at each plant. Also, a chemical plant at a site may be constructed to utilize more than one chemical process, for instance to produce multiple products.

A chemical plant commonly has usually large vessels or sections called units or lines that are interconnected by piping or other material-moving equipment which can carry streams of material. Such material streams can include fluids (gas or liquid carried in piping) or sometimes solids or mixtures such as slurries. An overall chemical process is commonly made up of steps called unit operations which occur in the individual units. A raw material going into a chemical process or plant as input to be converted into a product is commonly called a feedstock, or simply feed. In addition to feedstocks for the plant as a whole, an input stream of material to be processed in a particular unit can similarly be considered feed for that unit. Output streams from the plant as a whole are final products and sometimes output streams from individual units may be considered intermediate products for their units. However, final products from one plant may be intermediate chemicals used as feedstock in another plant for further processing. For example, some products from an oil refinery may used as feedstock in petrochemical plants, which may in turn produce feedstocks for pharmaceutical plants.

Either the feedstock(s), the product(s), or both may be individual compounds or mixtures. It is often not worthwhile separating the components in these mixtures completely; specific levels of purity depend on product requirements and process economics.

Operations

Chemical processes may be run in continuous or batch operation.

Batch Operation

In batch operation, production occurs in time-sequential steps in discrete batches. A batch of feedstock(s) is fed (or charged) into a process or unit, then the chemical process takes place, then the product(s) and any other outputs are removed. Such batch production may be repeated over again and again with new batches of feedstock. Batch operation is commonly used in smaller scale plants such as pharmaceutical or specialty chemicals production, for purposes of improved traceability as well as flexibility. Continuous plants are usually used to manufacture commodity or petrochemicals while batch plants are more common in speciality and fine chemical production as well as pharmaceutical active ingredient (API) manufacture.

Continuous Operation

In continuous operation, all steps are ongoing continuously in time. During usual continuous operation, the feeding and product removal are ongoing streams of moving material, which together with the process itself, all take place simultaneously and continuously. Chemical plants or units in continuous operation are usually in a steady state or approximate steady state. Steady state means that quantities related to the process do not change as time passes during operation. Such constant quantities include stream flow rates, heating or cooling rates, temperatures, pressures, and chemical compositions at any given point (location). Continuous operation is more efficient in many large scale operations like petroleum refineries. It is possible for some units to operate continuously and others be in batch operation in a chemical plant; for example, see Continuous distillation and Batch distillation. The amount of primary feedstock or product per unit of time which a plant or unit can process is referred to as the capacity of that plant or unit. For examples: the capacity of an oil refinery may be given in terms of barrels of crude oil refined per day; alternatively chemical plant capacity may be given in tons of product produced per day. In actual daily operation, a plant (or unit) will operate at a percentage of its full capacity. Engineers typically assume 90% operating time for plants which work primarily with fluids, and 80% uptime for plants which primarily work with solids.

Units and Fluid Systems

Specific unit operations are conducted in specific kinds of units. Although some units may operate at ambient temperature or pressure, many units operate at higher or lower temperatures or pressures. Vessels in chemical plants are often cylindrical with rounded ends, a shape which can be suited to hold either high pressure or vacuum. Chemical reactions can convert certain kinds of compounds into other compounds in chemical reactors. Chemical reactors may be packed beds and may have solid heterogeneous catalysts which stay in the reactors as fluids move through, or may be simply be stirred vessels in which reactions occur. Since the surface of solid heterogeneous catalysts may sometimes become "poisoned" from deposits such as coke, regeneration of catalysts may be necessary. Fluidized beds may also be used in some cases to ensure good mixing. There can also be units (or subunits) for mixing (including dissolving), separation, heating, cooling, or some combination of these. For example, chemical reactors often have stirring for mixing and heating or cooling to maintain temperature. When designing plants on a large scale, heat produced or absorbed by chemical reactions must be considered. Some plants may have units with organism cultures for biochemical processes such as fermentation or enzyme production.

Distillation unit in Italy

Separation processes include filtration, settling (sedimentation), extraction or leaching, distillation, recrystallization or precipitation (followed by filtration or settling), reverse osmosis, drying, and adsorption. Heat exchangers are often used for heating or cooling, including boiling or condensation, often in conjunction with other units such as distillation towers. There may also be storage tanks for storing feedstock, intermediate or final products, or waste. Storage tanks commonly have level indicators to show how full they are. There may be structures holding or supporting sometimes massive units and their associated equipment. There are often stairs, ladders, or other steps for personnel to reach points in the units for sampling, inspection, or maintenance. An area of a plant or facility with numerous storage tanks is sometimes called a *tank farm*, especially at an oil depot.

Fluid systems for carrying liquids and gases include piping and tubing of various diameter sizes, various types of valves for controlling or stopping flow, pumps for moving or pressurizing liquid, and compressors for pressurizing or moving gases. Vessels, piping, tubing, and sometimes other equipment at high or very low temperature are commonly covered with insulation for personnel safety and to maintain temperature inside. Fluid systems and units commonly have instrumentation such as temperature and pressure sensors and flow measuring devices at select locations in a plant. Online analyzers for chemical or physical property analysis have become more common. Solvents can sometimes be used to dissolve reactants or materials such as solids for extraction or leaching, to provide a suitable medium for certain chemical reactions to run, or so they can otherwise be treated as fluids.

Chemical Plant Design

Flow diagram for a typical oil refinery

Today, the fundamental aspects of designing chemical plants are done by chemical engineers. Historically, this was not always the case and many chemical plants were constructed in a haphazard way before the discipline of chemical engineering became established. Chemical engineering was first established as a profession in the United Kingdom when the first chemical engineering course was given at the University of Manchester in 1887 by George E. Davis in the form of twelve lectures covering various aspects of industrial chemical practice. As a consequence George E. Davis is regarded as the World's first Chemical Engineer.Today Chemical Engineering is a profession and those Professional Chemical Engineers with experience can gain "Chartered" engineer status through the Institution of Chemical Engineers.

In plant design, typically less than 1 per cent of ideas for new designs ever become commercialized. During this solution process, typically, cost studies are used as an initial screening to eliminate unprofitable designs. If a process appears profitable, then other factors are considered, such as safety, environmental constraints, controllability, etc. The general goal in plant design, is to construct or synthesize "optimum designs" in the neighborhood of the desired constraints.

Many times chemists research chemical reactions or other chemical principles in a laboratory, commonly on a small scale in a "batch-type" experiment. Chemistry information obtained is then used by chemical engineers, along with expertise of their own, to convert to a chemical process and scale up the batch size or capacity. Commonly, a small chemical plant called a pilot plant is built to provide design and operating information before construction of a large plant. From data and operating experience obtained from the pilot plant, a scaled-up plant can be designed for higher or full capacity.

After the fundamental aspects of a plant design are determined, mechanical or electrical engineers may become involved with mechanical or electrical details, respectively. Structural engineers may become involved in the plant design to ensure the structures can support the weight of the units, piping, and other equipment.

The units, streams, and fluid systems of chemical plants or processes can be represented by block flow diagrams which are very simplified diagrams, or process flow diagrams which are somewhat more detailed. The streams and other piping are shown as lines with arrow heads showing usual direction of material flow. In block diagrams, units are often simply shown as blocks. Process flow diagrams may use more detailed symbols and show pumps, compressors, and major valves. Likely values or ranges of material flow rates for the various streams are determined based on desired plant capacity using material balance calculations. Energy balances are also done based on heats of reaction, heat capacities, expected temperatures and pressures at various points to calculate amounts of heating and cooling needed in various places and to size heat exchangers. Chemical plant design can be shown in fuller detail in a piping and instrumentation diagram (P&ID) which shows all piping, tubing, valves, and instrumentation, typically with special symbols. Showing a full plant is often complicated in a P&ID, so often only individual units or specific fluid systems are shown in a single P&ID.

In the plant design, the units are sized for the maximum capacity each may have to handle. Similarly, sizes for pipes, pumps, compressors, and associated equipment are chosen for the flow capacity they have to handle. Utility systems such as electric power and water supply should also be included in the plant design. Additional piping lines for non-routine or alternate operating procedures, such as plant or unit startups and shutdowns, may have to be included. Fluid systems design commonly includes isolation valves around various units or parts of a plant so that a section of a plant could be isolated in case of a problem such as a leak in a unit. If pneumatically or hydraulically actuated valves are used, a system of pressurizing lines to the actuators is needed. Any points where process samples may have to be taken should have sampling lines, valves, and access to them included in the detailed design. If necessary, provisions should be made for reducing high pressure or temperature of a sampling stream, such including a pressure reducing valve or sample cooler.

Units and fluid systems in the plant including all vessels, piping, tubing, valves, pumps, compressors, and other equipment must be rated or designed to be able to withstand the entire range of pressures, temperatures, and other conditions which they could possibly encounter, including any appropriate safety factors. All such units and equipment should also be checked for materials compatibility to ensure they can withstand long-term exposure to the chemicals they will come in contact with. Any closed system in a plant which has a means of pressurizing possibly beyond the rating of its equipment, such as heating, exothermic reactions, or certain pumps or compressors, should have an appropriately sized pressure relief valve included to prevent overpressurization for safety. Frequently all of these parameters (temperatures, pressures, flow, etc.)

are exhaustively analyzed in combination through a *Hazop* or *fault tree analysis*, to ensure that the plant has no known risk of serious hazard.

Within any constraints the plant is subject to, design parameters are optimized for good economic performance while ensuring safety and welfare of personnel and the surrounding community. For flexibility, a plant may be designed to operate in a range around some optimal design parameters in case feedstock or economic conditions change and re-optimization is desirable. In more modern times, computer simulations or other computer calculations have been used to help in chemical plant design or optimization.

Plant Operation

Process Control

In process control, information gathered automatically from various sensors or other devices in the plant is used to control various equipment for running the plant, thereby controlling operation of the plant. Instruments receiving such information signals and sending out control signals to perform this function automatically are process *controllers*. Previously, pneumatic controls were sometimes used. Electrical controls are now common. A plant often has a control room with displays of parameters such as key temperatures, pressures, fluid flow rates and levels, operating positions of key valves, pumps and other equipment, etc. In addition, operators in the control room can control various aspects of the plant operation, often including overriding automatic control. Process control with a computer represents more modern technology. Based on possible changing feedstock composition, changing products requirements or economics, or other changes in constraints, operating conditions may be re-optimized to maximize profit.

Workers

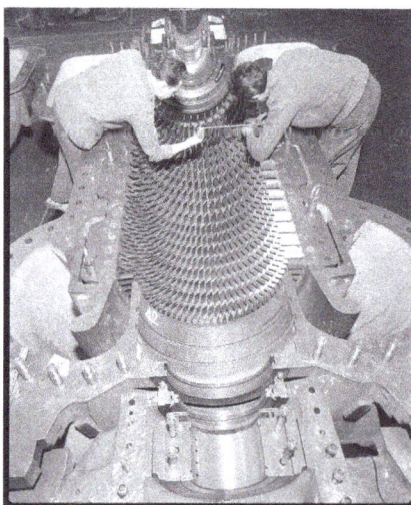

Workers in Italy, 1969. Photo by Paolo Monti

As in any industrial setting, there are a variety of workers working throughout a chemical plant facility, often organized into departments, sections, or other work groups. Such workers typically include engineers, plant operators, and maintenance technicians. Other personnel at the site could include chemists, management/administration and office workers. Types of engineers involved in operations or maintenance may include chemical process engineers, mechanical engineers for maintaining mechanical equipment, and electrical/computer engineers for electrical or computer equipment.

Transport

Large quantities of fluid feedstock or product may enter or leave a plant by pipeline, railroad tank car, or tanker truck. For example, petroleum commonly comes to a refinery by pipeline. Pipelines can also carry petrochemical feedstock from a refinery to a nearby petrochemical plant. Natural gas is a product which comes all the way from a natural gas processing plant to final consumers by pipeline or tubing. Large quantities of liquid feedstock are typically pumped into process units. Smaller quantities of feedstock or product may be shipped to or from a plant in drums. Use of drums about 55 gallons in capacity is common for packaging industrial quantities of chemicals. Smaller batches of feedstock may be added from drums or other containers to process units by workers.

Maintenance

In addition to feeding and operating the plant, and packaging or preparing the product for shipping, plant workers are needed for taking samples for routine and troubleshooting analysis and for performing routine and non-routine maintenance. Routine maintenance can include periodic inspections and replacement of worn catalyst, analyzer reagents, various sensors, or mechanical parts. Non-routine maintenance can include investigating problems and then fixing them, such as leaks, failure to meet feed or product specifications, mechanical failures of valves, pumps, compressors, sensors, etc.

Statutory and Regulatory Compliance

When working with chemicals, safety is a concern in order to avoid problems such as chemical accidents . In the United States, the law requires that employers provide workers working with chemicals with access to a Material Safety Data Sheet (MSDS) for every kind of chemical they work with. An MSDS for a certain chemical is prepared and provided by the supplier to whoever buys the chemical. Other laws covering chemical safety, hazardous waste, and pollution must be observed, including statutes such as the Resource Conservation and Recovery Act (RCRA) and the Toxic Substances Control Act (TSCA), and regulations such as the Chemical Facility Anti-Terrorism Standards in the United States. Hazmat (hazardous materials) teams are trained to deal with chemical leaks or spills. Process Hazard Analysis (PHA) is used to assess potential hazards in chemical plants. In 1998, the U. S. Chemical Safety and Hazard Investigation Board has become operational.

Plant Facilities

The actual production or process part of a plant may be indoors, outdoors, or a combination of the two. It may be a traditional stick-built plant or a modular skid. Large modular skids are especially impressive feats of engineering. A modular skid is built including all of the modular equipment needed to do the same job a traditional stick-build plant may perform. However, the modular skid is built within a structural steel frame, allowing it to be shipped to the onsite location without needing to be rebuilt onsite. A modular skid build results in a higher functioning end product, as less hands are required in the onsite setup of the modular skid process unit, resulting in minimized risk for mishaps. The actual production section of a facility usually has the appearance of a rather industrial environment. Hard hats and work shoes are commonly worn. Floors and stairs are often made of metal grating, and there is practically no decoration. There may also be pollution control or waste treatment facilities or equipment. Sometimes existing plants may be expanded or modified based on changing economics, feedstock, or product needs. As in other production facilities, there may be shipping and receiving, and storage facilities. In addition, there are usually certain other facilities, typically indoors, to support production at the site.

Although some simple sample analysis may be able to be done by operations technicians in the plant area, a chemical plant typically has a laboratory where chemists analyze samples taken from the plant. Such analysis can include chemical analysis or determination of physical properties. Sample analysis can include routine quality control on feedstock coming into the plant, intermediate and final products to ensure quality specifications are met. Non-routine samples may be taken and analyzed for investigating plant process problems also. A larger chemical company often has a research laboratory for developing and testing products and processes where there may be pilot plants, but such a laboratory may be located at a site separate from the production plants.

A plant may also have a workshop or maintenance facility for repairs or keeping maintenance equipment. There is also typically some office space for engineers, management or administration, and perhaps for receiving visitors. The decorum there is commonly more typical of an office environment.

Clustering of Commodity Chemical Plants

Chemical Plants used particularly for commodity chemical and petrochemical manufacture,are located in relatively few manufacturing locations around the world largely due to infrastructural needs.This is less important for speciality or fine chemical batch plants. Not all commodity/petrochemicals are produced in any one location but groups of related materials often are, to induce industrial symbiosis as well as material, energy and utility efficiency and other economies of scale. These manufacturing locations often have business clusters of units called chemical plants that share utilities and large scale infrastructure such as power stations, port facilities, road and rail terminals. In the United Kingdom for example there are four main locations for commodity chemical manufacture: near the River Mersey in Northwest England, on the Humber on the East coast of Yorkshire, in Grangemouth near the Firth of Forth in Scotland and on Teesside as part of the Northeast of England Process Industry Cluster (NEPIC). Approximately 50% of the UK's petrochemicals, which are also commodity chemicals, are produced by the industry cluster companies on Teesside at the mouth of the River Tees on three large chemical parks at Wilton, Billingham and Seal Sands.

Corrosion and use of New Materials

Corrosion in chemical process plants is a major issue that consumes billions of dollars yearly. Electrochemical corrosion of metals is pronounced in chemical process plants due to the presence of acid fumes and other electrolytic interactions. Recently, FRP (Fibre-reinforced plastic) is used as a material of construction. The British standard specification BS4994 is widely used for design and construction of the vessels, tanks, etc.

Cryogenic Nitrogen Plant

Nitrogen, as an element of great technical importance, can be produced in a cryogenic nitrogen plant with a purity of more than 99.9999%. Air inside a distillation column is separated at cryogenic temperatures (about 100K/-173°C) to produce high purity nitrogen with 1ppm of impurities. The process is based on the air separation, which was invented by Dr. Carl von Linde in 1895.

Purpose

Gasious Nitrogen (GAN) plant with production rate of 800 Nm³/hour

The main purpose of a cryogenic nitrogen plant is to provide a customer with high purity gaseous nitrogen (GAN). In addition liquid nitrogen (LIN) is produced simultaneously and is typically 10% of the gas production. High purity liquid nitrogen produced by cryogenic plants is stored in a local tank and used as a strategic reserve. This liquid can be vaporised to cover peaks in demand or for use when the nitrogen plant is offline. Typical cryogenic nitrogen plants range from 250 Nm³/hour to very large range plants with a daily capacity of 63.000 tonnes of nitrogen a day (as the Cantarell Field plant in Mexico).

Plant Modules

A cryogenic nitrogen plant comprises:

Warm End (W/E) Container

- Compressor
- Air receiver
- Chiller (Heat exchanger)
- Pre-filter
- Air purification unit (APU)

Coldbox

- Main heat exchanger
- Distillation Column
- Condenser
- Expansion brake turbine

Storage and Backup System

- Liquid nitrogen tank
- Vapouriser

How the Plant Works

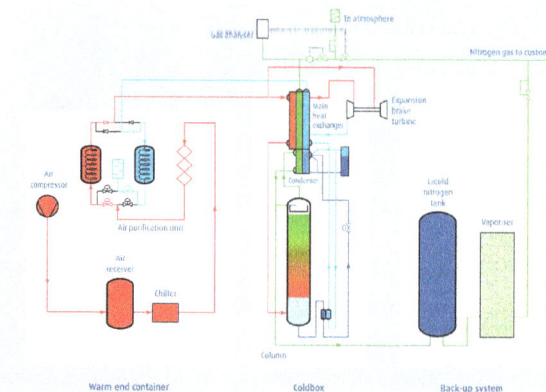

Flowsheet GAN Plant Linde Cryoplants Ltd.

Warm End Process

Atmospheric air is roughly filtered and pressurised by a compressor, which provides the product pressure to deliver to the customer. The amount of air sucked in depends on the customer's nitrogen demand.

The Air Receiver collects condensate and minimises pressure drop. The dry and compressed air leaves the air to refrigerant heat exchanger at about 10°C.

To clean the process air further, there are different stages of filtration. First of all, more condensate is removed, this removes some hydrocarbons.

The last unit process in the warm end container is the thermal swing adsorber (TSA). The Air purification unit cleans the compressed process air by removing any residual water vapour, carbon dioxide and hydrocarbons. It comprises two vessels, valves and exhaust to allow the changeover of vessels. While one of the TSA beds is on stream the second one is regenerated by the oxygen rich waste flow, which is vented through a silencer into the ambient environment.

Coldbox Process

After leaving the air purification unit, the process air enters the main heat exchanger, where it is rapidly cooled down to -165°C. All residual impurities (e.g. CO_2) freeze out, and the process air enters at the bottom of the distillation column partially liquefied.

Back up Process

Liquid Nitrogen produced from the cold box transfers into the liquid storage tank. An ambient air vaporiser is used to vaporise stored liquid during peak demand. A pressure control panel senses the demand for gaseous nitrogen and regulates the gas flow into the end-users pipeline to maintain line pressure.

Applications for High Purity Nitrogen Production

- Ammonia production for the fertilizer industry
- Float glass manufacture
- Petrochemical
    - o Purge gas
    - o Blanketing/Inerting gas for tanks and reactor vessels
    - o Amine gas treatment
    - o Bearing seal gas

- Polyester manufacture

- Semiconductor manufacture

- Photovoltaic manufacture

Cryogenic Oxygen Plant

A cryogenic oxygen plant is an industrial facility that creates molecular oxygen at relatively high purity. Oxygen is the most common element in the earth's crust and the second largest industrial gas. This process was pioneered by Dr. Carl von Linde in 1902.

Liquid Oxygen plant with production rate of 80 Nm³/hour

Purpose

The cryogenic air separation achieves high purity oxygen of more than 99.5%. The resulting high purity product can be stored as a liquid and/or filled into cylinders. These cylinders can even be distributed to customer in the medical sector, welding or mixed with other gases and used as breathing gas for diving. Typical production ranges from 50Nm³/hour up to 860.000Nm³/hour (Ras Laffan refinery).

Plant Modules

A cryogenic oxygen plant comprises:

Warm End (W/E) Container

- Compressor

- Air receiver

- Chiller (Heat exchanger)

- Pre-filter

- Air purification unit (APU)

Coldbox

- Main heat exchanger

- Boiler

- Distillation column

- Expansion brake turbine

Storage

- Liquid oxygen tank

- Vapouriser

- Filling station

How the Plant Works

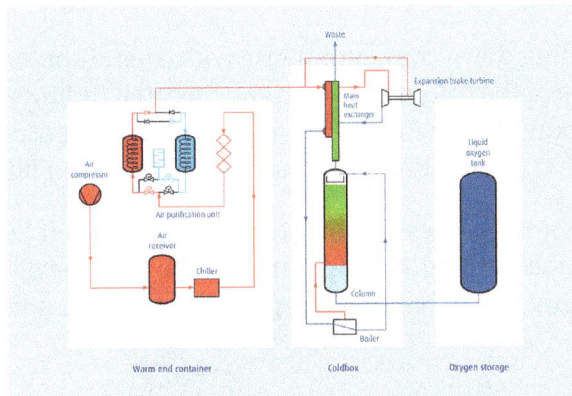

Flowsheet Liquid Oxygen (LOX) Plant

Warm End Process

Atmospheric air is roughly filtered and pressurised by a compressor, which provides the product pressure to deliver to the customer. The amount of air sucked in depends on the customer's oxygen demand.

The air receiver collects condensate and minimises pressure drop. The dry and compressed air leaves the air to refrigerant heat exchanger with about 10°C.

To clean the process air further, there are different stages of filtration. First of all, more condensate is removed, then a Coalescing filter acts as a gravity filter and finally an adsorber filled with activated carbon removes some hydrocarbons.

The last unit process in the warm end container is the thermal swing adsorber (TSA). The Air purification unit cleans the compressed process air by removing any residual water vapour, carbon dioxide and hydrocarbons. It comprises two vessels, valves and exhaust to allow the changeover of vessels. While one of the TSA beds is on stream the second one is regenerated by the waste gas flow, which is vented through a silencer into the ambient environment.

Coldbox Process

The process air enters the main heat exchanger in the coldbox where it is cooled in counter flow with the waste gas stream. After leaving the main heat exchanger the process air has a temperature of about –112°C and is partly liquefied. The complete liquefaction is achieved through evaporation of cooled liquid oxygen in the boiler. After passing a purity control valve process air enters on tip of the distillation column and flows down through the packing material.

The steam of evaporated oxygen vapour in the shell of the boiler vents back into the distillation column. It rises through the column packing material and encounters the descending stream of liquid process air.

The liquid air descending down the column loses nitrogen. It becomes richer in oxygen and collects at the base of the column as pure liquid oxygen. It flows out into the boiler to the cold box liquid product valve. An on-line oxygen analyser controls the opening of the liquid product valve to transfer pure low-pressure liquid oxygen into the storage tank.

The rising oxygen vapour becomes rich in nitrogen and argon. It leaves the column and exits the cold box at ambient temperature through the main heat exchanger as a waste gas. This waste gas provides purge gas to regenerate the TSA unit and to the cool the refrigeration turbine.

Turbines located at the base of the cold box provide refrigeration for the process. A stream of high-pressure gas from the main heat exchangers is cooled and expanded to low pressure in the turbine. This cold air returns to the waste stream of the heat exchanger to inject refrigeration. Energy removed by the turbine re-appears as heat in the turbine's closed-cycle air-brake circuit. This heat is removed in an air-to-air cooler by waste gas from the cold box.

Storage and Vaporising Process

Liquid from the tank is compressed to high pressure in a cryogenic liquid pump. It is then vaporised in an ambient air evaporator to produce gaseous oxygen. The high-pressure gas then can pass into cylinders via the gas manifold or fed into a customers product pipeline.

Applications

- Furnace enrichment

- Medical gases

- Metal production

- Welding

Unit Operation

In chemical engineering and related fields, a unit operation is a basic step in a process. Unit operations involve a physical change or chemical transformation such as separation, crystallization, evaporation, filtration, polymerization, isomerization, and other reactions. For example, in milk processing, homogenization, pasteurization, chilling, and packaging are each unit operations which are connected to create the overall process. A process may require many unit operations to obtain the desired product from the starting materials, or feedstocks.

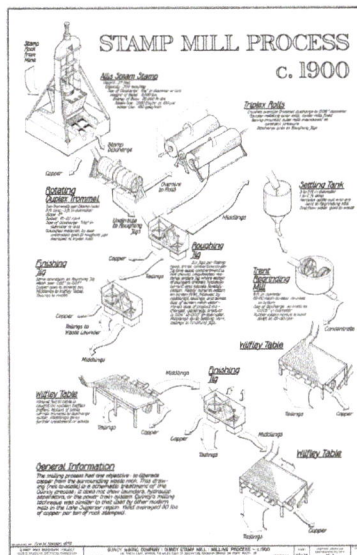

An ore extraction process broken into its constituent unit operations
(Quincy Mine, Hancock, MI ca. 1900)

History

Historically, the different chemical industries were regarded as different industrial processes and with different principles. Arthur Dehon Little propounded the concept of "unit operations" to explain industrial chemistry processes in 1916. In 1923, William

H. Walker, Warren K. Lewis and William H. McAdams wrote the book *The Principles of Chemical Engineering* and explained that the variety of chemical industries have processes which follow the same physical laws. They summed up these similar processes into unit operations. Each unit operation follows the same physical laws and may be used in all relevant chemical industries. For instance, the same engineering is required to design a mixer for either napalm or porridge, even if the use, market or manufacturers are very different. The unit operations form the fundamental principles of chemical engineering.

Chemical Engineering

Chemical engineering unit operations consist of five classes:

1. Fluid flow processes, including fluids transportation, filtration, and solids fluidization.

2. Heat transfer processes, including evaporation and heat exchange.

3. Mass transfer processes, including gas absorption, distillation, extraction, adsorption, and drying.

4. Thermodynamic processes, including gas liquefaction, and refrigeration.

5. Mechanical processes, including solids transportation, crushing and pulverization, and screening and sieving.

Chemical engineering unit operations also fall in the following categories which involve elements from more than one class:

*   Combination (mixing)

*   Separation (distillation, crystallization)

*   Reaction (chemical reaction)

Furthermore, there are some unit operations which combine even these categories, such as reactive distillation and stirred tank reactors. A "pure" unit operation is a physical transport process, while a mixed chemical/physical process requires modeling both the physical transport, such as diffusion, *and* the chemical reaction. This is usually necessary for designing catalytic reactions, and is considered a separate discipline, termed chemical reaction engineering.

Chemical engineering unit operations and chemical engineering unit processing form the main principles of all kinds of chemical industries and are the foundation of designs of chemical plants, factories, and equipment used.

In general, unit operations are designed by writing down the balances for the transported quantity for each elementary component (which may be infinitesimal) in the form of equations, and solving the equations for the design parameters, then selecting an optimal solution out of the several possible and then designing the physical equipment. For instance, distillation in a plate column is analyzed by writing down the mass balances for each plate, wherein the known vapor-liquid equilibrium and efficiency, drip out and drip in comprise the total mass flows, with a sub-flow for each component. Combining a stack of these gives the system of equations for the whole column. There is a range of solutions, because a higher reflux ratio enables fewer plates, and vice versa. The engineer must then find the optimal solution with respect to acceptable volume holdup, column height and cost of construction.

Control of Distillation Systems

Distillation Basics

Distillation is unarguably the most preferred unit operation used for separating mixtures. In the design of chemical processes, other separation techniques are considered only if distillation is found to be economically unviable. It is thus not surprising that the final product stream from a plant is typically a product steam from a distillation column. This Chapter provides guidelines for designing effective control systems for distillation columns.

The Simple Distillation Column

A proper understanding of the basic physics of a distillation column (or any other process for that matter) is a pre-requisite for designing an effective control system. Figure shows the schematic of a simple distillation column along with the control valves. It consists of a tray section, a condenser, a reflux drum and a reboiler. The feed mixture is fed on a feed tray. The trays above the feed tray constitute the rectifying / enriching section and those below constitute the stripping section. The overhead distillate and the bottoms are the two product streams from a simple distillation column. Steam is typically used to provide vapour reboil into the stripping section. The liquid reflux into the enriching section is provided by the condenser. Cooling water is commonly used as the coolant in the condenser. The condenser may be a total condenser, where all the vapour is condensed, or a partial condenser where only a part of the vapour is condensed. The overhead distillate is a liquid stream for a total condenser. A partial condenser column may be operated at total reflux where all the liquid is refluxed back into the column and the distillate stream is a vapour stream. Alternatively (and more commonly) both a vapour and a liquid distillate stream are drawn. The reflux drum provides surge capacity to adjust the reflux and distillate rate during transients. The bottom sump provides the surge capacity for adjusting the bottoms and steam rate.

The vapour generated when a volatile liquid feed mixture is boiled is richer in the more volatile component. The remaining liquid is then richer in the heavier components. Chemical engineers refer to this as flashing a mixture. If the flashed vapour is condensed and partially vaporized again, the vapour from the second flash would be further enriched in the volatiles (light boilers). Similarly, if the liquid from the first flash is further vaporized, the heavies composition of the liquid from the second flash would increase. Theoretically speaking, a sufficiently large number of flash operations on the vapour can result in a final vapour stream that is almost 100% pure lightest component. Similarly a series of flash operations on the liquid can result in a final liquid product that is 100% heaviest component. The array of trays in a distillation column accomplishes this series of flash operations. The temperature difference between the liquid and vapour streams entering a tray causes condensation / vaporization so that as one moves up the column, the composition of the lightest component increases monotonically. Alternatively, as one moves down the column, the composition of the heaviest component keeps on increasing. Since heavier components boil at higher temperatures, the tray temperature increases as one moves down the column with the condenser being the coolest and the reboiler being the hottest. The reboiler and the condenser are the source of vaporization and condensation respectively for the series of vaporization / condensation.

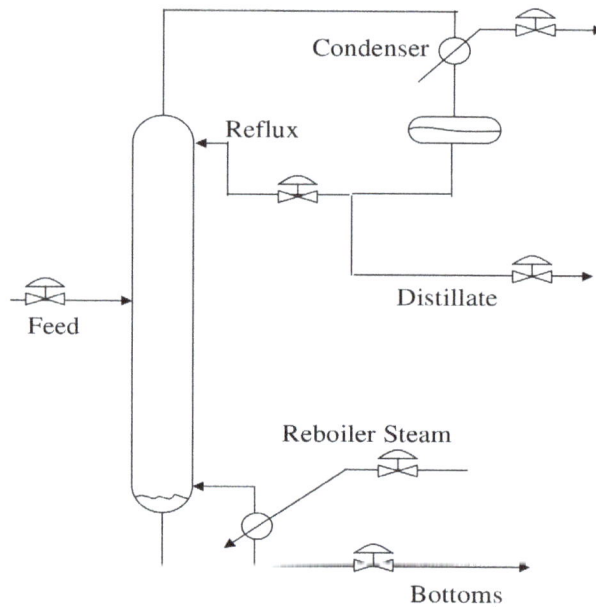

Schematic of a simple distillation column along with the control valves.

Splits in a Simple Distillation Column

Consider a five component equimolar ABCDE mixture feed into a simple distillation column. The components are in decreasing order of volatility so that A is the lightest and E is the heaviest. The feed rate is 100 kmol/h. The steady state distillate to bottoms product split is primarily determined by the choice of the distillate (or bottoms) rate. Assuming a sufficiently large number of trays, adequate reboil and reflux, for a distillate rate of 40 kmol / hr, which is equal to the component A and component B flow rate in the feed, essentially all of the A and B would leave up the top so that the distillate would contain traces C, D and E impurities in decreasing order of composition. The bottoms would be a CDE mixture with traces of B and A, in decreasing order of composition. The column thus accomplishes a split between components B and C with the liquid preventing C from escaping up the top and the vapour reboil preventing B from escaping down the bottoms. Components B and C, are referred to as the light key (LK) and heavy key (HK) respectively. The LK is the dominant impurity in the bottoms stream and the HK is the dominant impurity in the distillate stream. The component split is referred to as an AB/CDE split. The component that is the next lighter component than the LK is called the lighter than light key (LLK). The heavier than heavy key (HHK) can be defined in a complementary manner. Components A and E are respectively the lightest and heaviest and therefore referred to as the lightest key and the heaviest key.

For the ABCDE mixture, there are four possible splits – A/BCDE, AB/CDE, ABC/DE and ABCD/E. The first one, where the light key is also the lightest key is referred to as the direct split. The last one, where the heavy key is also the heaviest key is referred to as an indirect split. The remaining splits where the key components are intermediate boilers are referred to as intermediate splits. It is helpful to categorize the column split into these basic types.

Basic Control Structures

A simple distillation column with a total condenser has a total of six valves as in Figure. Of these six valves, the feed valve is usually set by an upstream unit in the process. Also two valves must be used to control the reflux drum level and the reboiler level as liquid levels are non-self regulating. Another valve must be used to regulate the column pressure which represents the vapour inventory in the column. Typically, the cooling duty valve in the condenser is used for pressure control. After implementing the three inventory loops, the position of the remaining two control valves can be set by an operator or a controller to regulate the separation. This gives a operation degree of freedom of two for a simple distillation column. The operation degree of freedom is more for complex column configurations that are considered later.

Four control structure types result for a distillation column corresponding to the choice of valve used for reflux drum and reboiler level control. These are the LQ, DQ, LB and DB structures and are illustrated in Figure. The nomenclature corresponds to the two control degrees of freedom (valves) that remain to regulate the separation. The LQ control structure corresponds to the distillate (D) controlling the reflux drum level and the bottoms (B) controlling the reboiler level. This leaves the reflux (L) and reboiler duty (Q) as the two valves for regulating the separation achieved, hence the label LQ. In the DQ structure, the condenser level is controlled using the reflux while in the LB structure, the bottoms level is controlled using the reboiler duty. Lastly in the DB control structure, the reboiler duty and reflux are used for controlling the reboiler and condenser levels respectively.

The Energy Balance (LQ) Structure

The LQ control structure is the most natural control structure for a simple distillation column. This is because the separation in a distillation column occurs due to successive condensation and vaporization of the counter-current vapour and liquid streams flowing through the column. Adjusting the cold reflux, the source of condensation, and the reboiler duty, the source of vaporization, is then a natural choice for regulating the separation achieved in the column. The LQ control structure is thus the most commonly applied distillation control structure. It is also sometimes referred to as an energy balance structure as changing L (cold reflux) or Q alters the energy balance across the column to affect the distillate to bottoms product split.

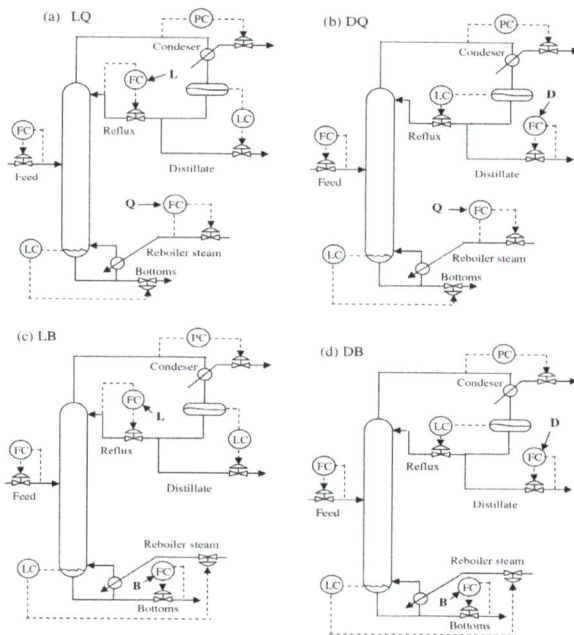

Schematics of LQ, DQ, LB and DB control structures

Material Balance Structures

The other control structures are referred to as material balance structures as the product split is directly adjusted by changing the distillate or bottoms stream flow rate. The material balance structures are applied when a level loop for the LQ structure would be ineffective due to a very small product stream (D or B) flow rate. The DQ structure is thus appropriate for columns with very large reflux ratio (L/D > 4). The distillate stream flow is then a fraction of the reflux stream so that the reflux drum level cannot be maintained using the distillate. The level must then be controlled using the reflux. The LB structure is appropriate for columns with a small bottoms flow rate compared to the boil-up. The bottoms stream is then not appropriate for level control and the reboiler duty must be used instead. The DB control structure is used very rarely as both D and B cannot be set independently due to the steady state overall material balance constraint. In dynamics however, the control structure may be used when the reflux and reboil are much larger than the distillate and bottoms respectively.

Other Control Structure Variants

Other variants of the basic control structure types include the L/D-Q, L/D-B and D-Q/B. In the first two structures the reflux ratio is adjusted for regulating the separation. In the last structure the reboil ratio is adjusted. These control structures are illustrated in Figure. Note that when the reflux is adjusted in ratio with the distillate, the distillate stream can be used to control the reflux drum level even as it may be a trickle compared to the reflux rate. Similarly, when the reboil rate is adjusted in ratio with the bottoms, a small bottoms stream can provide effective level control.

Maintaining the reflux ratio is quite common in distillation control as it provides feedforward compensation in the reflux for a change in the distillate rate. Such a feedforward compensation can significantly improve quality control as the column dynamics are slow with respect to a change in the reflux rate due to the slow liquid hydraulics with every tray having a time constant of 15-30 s. Pure feedback adjustment of the reflux can thus result in large purity deviations. Maintaining the reboil ratio is not very popular. This is because all the tray compositions / tray temperatures respond almost immediately to a change in the reboil due to the fast vapour dynamics. Adjustment of the reboiler duty in a feedback arrangement is thus usually sufficient for effective regulation.

Temperature Based Inferential Control

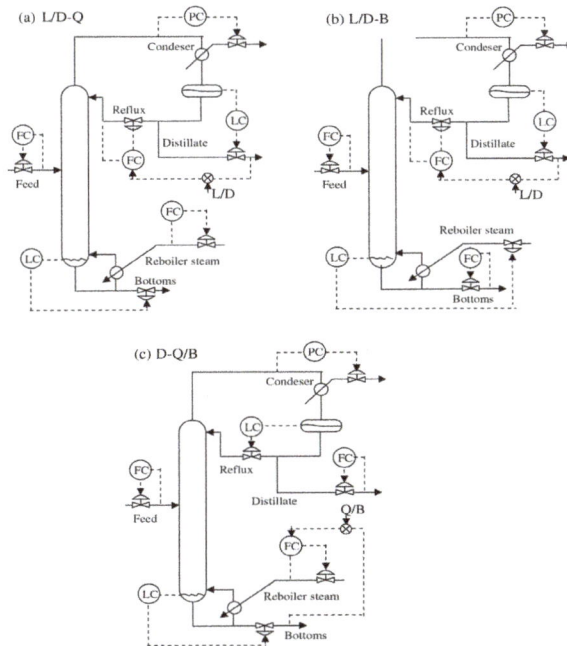

Schematics of L/D-Q, L/D-B, and D-Q/B control structures.

The distillation column performs a separation between the light key and the heavy key so that heavy key and light key impurity levels respectively in the distillate and bottoms are below design specifications. The primary control objective then is to ensure these impurity levels remain below specifications for load changes. A change in the flow rate and composition of the feed into the column are the two major load disturbances that must be rejected by the control system. Feedback control based on the impurity levels in the product streams is usually not acceptable due to the large delays / lags introduced by composition measurements. Also, control action would only be taken after the product purity deviates in a feedback system. Considering that the column consists of an array of trays, the trays would respond to a load disturbance much before the effect of the disturbance reaches the product streams. It therefore makes sense to control an appropriate tray process variable so that the disturbance is compensated for before the product purities are affected. This would lead to tighter product purity control.

The tray temperature is almost always used as an inferential variable for the tray composition. The boiling point of a mixture depends on the pressure and the mixture composition. At a constant pressure, the boiling point increases as the concentration of heavies increases and vice versa. A change in the tray composition can thus be inferred from a change in the tray temperature. The relationship is exact for a binary mixture and is approximate for a multi-component mixture.

Single-Ended Temperature Control

Controlling a single tray temperature in a column is usually referred to as single-ended temperature control. Either of the two operation degrees of freedom can be used as the manipulation handle. For example, in the LQ control structure, the reflux rate or the reboiler duty can be manipulated for maintaining a tray temperature. This is shown in Figure (a). Manipulation of Q is usually preferred due to the fast response of all the tray temperatures to a change in Q. The dynamics with respect to reflux rate are slower due to the associated tray liquid hydraulic lags. The reflux rate can also be used for maintaining a tray temperature if the control tray location is not too far below the reflux (say about 10 trays). The single ended variants for the DQ and LB control structures are shown in Figure (b) and (c) respectively. In the DQ structure, if the distillate rate is used to control a tray temperature, the temperature controller is nested with the reflux drum level controller. This means that the temperature controller would work only if the reflux drum level controller is working. Similarly, in the LB structure, if the bottoms flow rate controls a tray temperature, the temperature controller is nested with the reboiler level controller. In both these cases, the level controller must be a tightly tuned PI controller, else the temperature control would be extremely sluggish. Note that the reflux and the reboil are the only two causal variables that affect the tray temperature so that any control scheme must directly / indirectly effect a change in these causal variables.

Dual-Ended Temperature Control

Theoretically speaking, since the column degree of freedom is two, two tray temperatures can be controlled in a column. This is referred to as dual-ended temperature control. For example, in the LQ control structure, the reflux rate can be used for controlling a rectifying tray temperature and the reboiler duty can be used to control a stripping tray temperature as in Figure. Industrial practice is to control a single tray temperature as controlling two tray temperatures usually requires detuning of the temperature controllers due to interaction between the temperature loops. More importantly, the two controlled tray temperatures may not be sufficiently independent so that, in the worst case, the control system may seek infeasible temperature set-points. Dual temperature control is possible for long towers so that two tray temperatures that are far apart are sufficiently independent.

Temperature Sensor Location Selection

Various criteria have evolved for the selection of the most appropriate tray location(s) for temperature control. Prominent among these are selection of tray with the maximum slope in the temperature profile, sensitivity analysis and SVD analysis.

Maximum Slope Criterion

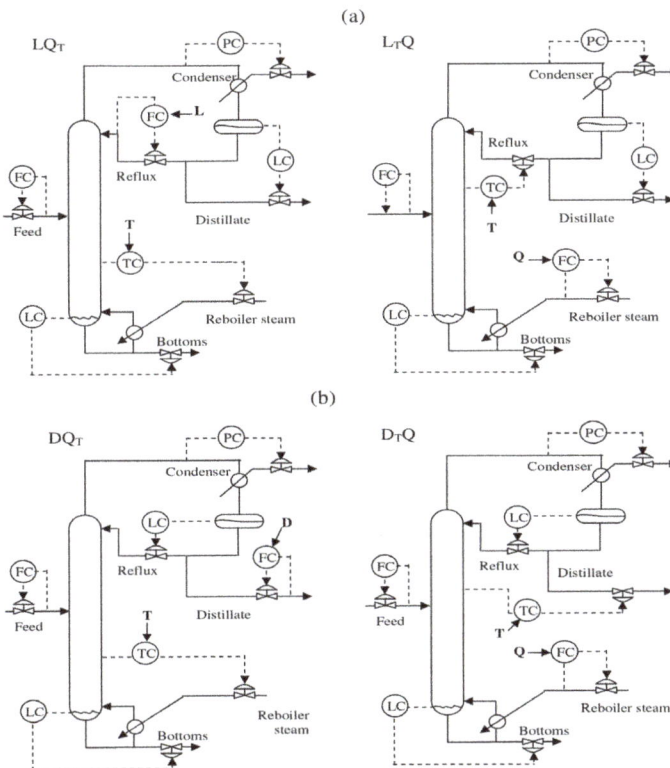

Single ended temperature control structures using LQ and DQ scheme

The maximum slope criterion is the simplest to use and requires only the steady state temperature profile. From the temperature profile, the tray location where the temperature changes the most from one tray to the other is chosen as the control tray. The temperature profile usually also shows a large change / break at the feed tray location. The feed tray should however not be chosen for control as the changes in temperature would be due to changes in the feed composition / temperature and not due to a change in the separation. A large change in the temperature from one tray to the other reflects large separation between the key components so that disturbances in the separation would affect this separation zone much more than other locations. It therefore makes sense to place the temperature sensor at that location.

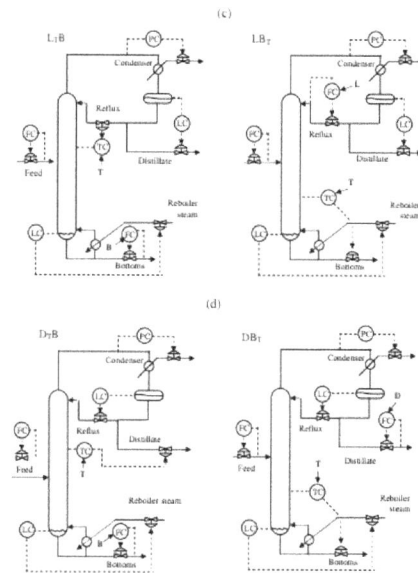

Single ended temperature control structures using LB and DB scheme

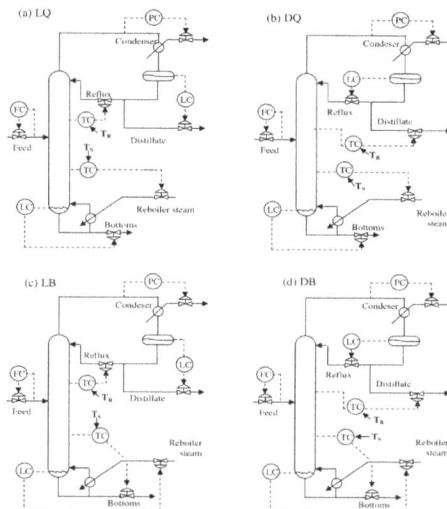

Dual ended temperature control structures using LQ, DQ, LB and DB schemes.

Maximum Sensitivity Criterion

Sensitivity analysis recommends controlling the tray with maximum sensitivity to the control input. The causal variables that effect a change in the tray temperature are the reflux rate (or ratio) and the reboiler duty. The sensitivity of the ith tray temperature to the reflux rate (L) and reboiler duty (Q) is defined as

$$S_{iL} = \frac{\partial T_i}{\partial L}$$

And $\qquad S_{iQ} = \frac{\partial T_i}{\partial Q}$

respectively. Controlling the most sensitive tray location provides muscle to the controller as a smaller change in the manipulated variable is needed to bring the deviating temperature back to its set-point. The open loop steady state gain is large so that a low controller gain suffices. The low controller gain mitigates sensor noise amplification. Also, a small bias in the temperature sensor can be tolerated. Plotting the sensitivity of all the trays with respect to Q and L would reveal the most sensitive tray location. In case two distinct regions of high sensitivity are observed, dual temperature control should be possible. If not, dual temperature control is likely to result in the two temperature controllers fighting each other.

SVD Criterion

The SVD analysis is another useful technique for selecting the tray temperature locations. The sensitivity matrix

$$S = \begin{bmatrix} S_L & S_Q \end{bmatrix}$$

where S_L and S_Q are column vectors of tray sensitivities, is decomposed using the singular value decomposition (SVD) as

$$S = U \Sigma V^T$$

In the above U and V are orthogonal matrices with the columns constituting the left singular and right singular vectors, respectively. The Σ matrix is a diagonal matrix. A plot of the first two left singular vectors (first two columns of U) shows the two most independent locations in the column. The index of the element of the first left singular with the maximum magnitude corresponds to the tray location that should be controlled in a single-ended scheme. If dual temperature control is to be implemented, the corresponding index for the second left singular vector gives the tray location for the second temperature sensor. The feasibility of dual ended temperature control is reflected in the ratio of the two diagonal elements, σ_1 and σ_2, of Σ. The diagonal elements in Σ are always in decreasing order of magnitude. If the two singular values are comparable, ie the ratio σ_1 / σ_2 is not too large (say < 10), dual temperature control should be possible.

Of the above three criteria, the maximum slope criteria is the simplest to use. Sensitivity analysis and SVD analysis requires the availability of a rating program to calculate the tray temperature sensitivities to the manipulated variables. The SVD technique further requires a module to obtain the SVD decomposition of the sensitivity matrix S. In most distillation column studies, the three techniques would agree on sensor location. However, for columns with highly non-ideal columns, the use of the SVD technique is recommended for selecting the tray location.

Considerations in Temperature Inferential Control

Effect of LLK / HHK

The temperature composition relationship is not exact for multi-component mixtures. If the feed LLK composition increases, the LLK must leave up the top of the column. For the same LK/HK split, the tray temperatures in the enriching section should be lower as the LLK composition must increase due to the increase in its feed composition. This dip in the temperature would be more as one moves up the enriching section since the LLK accumulates at the top. If a tray temperature near the top is controlled using the reboiler duty and the tray temperature set-point is not reduced on increase of feed LLK, more of the HK would be pushed up the top of the column by the action of the controller. Controlling a tray temperature near the feed tray would mitigate this effect. Another option is to measure the HK composition in distillate and use it to compensate the control tray temperature set-point. Note that the volatilities dictate that any LLK entering the column must exit up the top. In case LLK in the distillate is not acceptable, action must be taken upstream to ensure LLKs do not enter the column. Troubleshooting the process would typically reveal an upstream column not doing its job.

Similar to LLK, if a stripper tray temperature low down the column is being controlled using boilup, an increase in feed HHK would cause more LK to leak down the bottoms, unless the tray temperature setpoint is appropriately increased.

Flat Temperature Profiles

When the key components in a mixture are close boiling, the column temperature profile is flat with only a small change in adjacent tray temperatures. This is typical of superfractionators that use a large number of trays and a high reflux ratio as the separation is inherently difficult. Controlling a tray temperature is then not desirable as variations in the tray pressure with changes in column internal flow rates would swamp any subtle variations in the tray temperature due to composition changes. Controlling the difference in two tray temperatures that are located close by mitigates the effect of pressure variation as the change in the local pressure for the two trays would be about the same. The differential temperature measurement then reflects the change in the HK (or LK) composition between the trays. Care must be exercised in the use of a differential temperature measurement as the variation in ΔT with the bottoms composition depends on the location of the separation zone inside the column. If the measurement trays are below the separation zone, ΔT increases as the steam rate is decreases. Once the separation zone passes below the ΔT trays, a decrease in the steam would cause the ΔT to decrease. The gain thus changes sign depending on the location of the separation zone inside the column.

Easy Separations

The other extreme to a flat temperature profile is an extremely sharp temperature profile. This occurs when the separation is very easy so that the separation zone shows a large change in temperature over a few trays ie a sharp temperature profile. During transients, this sharp separation zone may move up or down the column leading to temperature transmitter saturation. Once the separation zone moves up and continues to move up, the error signal that the controller sees does not change so that the burden of bringing the profile back falls on the integral action. The problem is compounded by the low controller gain due to the extreme sensitivity of the tray temperature to a change in the manipulated variable. The problem is solved by controlling the average temperature of the trays over which the profile moves.

Control of Complex Column Configurations

Side-draw Columns

Side product streams are sometimes withdrawn from a column when the product purity specifications are not very tight and there is small amount of impurity in the feed that must be purged. Two common configurations are a liquid side-draw from a tray above the feed tray or a vapour side draw below the feed tray. Consider an ABC ternary mixture. If the component flow rate of A in the fresh feed is small, the liquid side stream withdrawal above the feed tray allows most of the B to be removed in the side stream. The side-draw must be liquid as A being the LLK would be present in smaller amounts in the liquid phase. The vapour side draw below the feed tray is used when there is a small amount of C (compared to A and B) in the fresh feed. The C HHK would separate into the liquid phase so that a vapour side stream that is mostly B with small amounts of C can be withdrawn below the feed tray. The side stream (liquid or vapour) provides an additional opeoration degree of freedom and its flow rate may be adjusted to maintain the B purity in the side draw. The control schemes are illustrated in Figure.

Alternative simpler control schemes are possible when the light A or heavy B impurities occur in very small amounts in the fresh feed. The purge rate is flow controlled with a set-point corresponding to the maximum expected impurity component flow in the feed. When the impurity is below this maximum, small amount of LK or HK would be lost with purge. However, the loss is acceptable due to the very small purge rate. The alternative simpler control schemes for the two common side draw configurations are shown in Figure.

Side Rectifier / Side Stripper Columns

The side rectifier and side stripper columns are an extension of the side-draw column discussed above. As with side-stream columns, these are used when there is a small amount of light or heavy impurity in the feed that is removed as a small purge stream. However the purity specs on the main products are tight. The vapour or liquid side stream respectively, must then be further rectified or stripped to ensure that the impurity is pushed back into the main column and does not escape with side-product stream to ensure high purity. The side stripper and side rectifier column arrangements are shown in Figure (a) and (b). An additional operation degree of freedom is introduced in the form of the reflux rate or the reboiler duty. The side draw rate and the reflux rate or reboiler duty can then be adjusted to maintain the two impurities in the side-product. The corresponding control schemes are shown in Figure (a). Along with the two composition loops in the main column, these schemes represent a highly coupled 4X4 multivariable system. Simpler control schemes with only one temperature (or composition) being controlled in each of the main column and the side-column are much more practical.

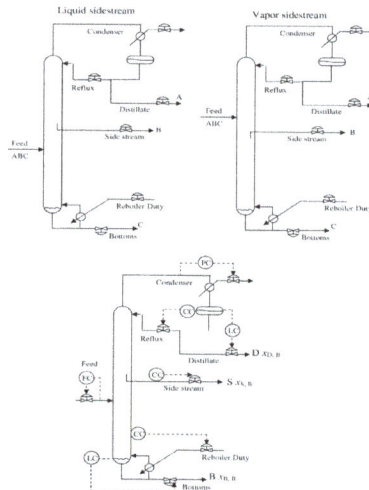

Control of side stream column

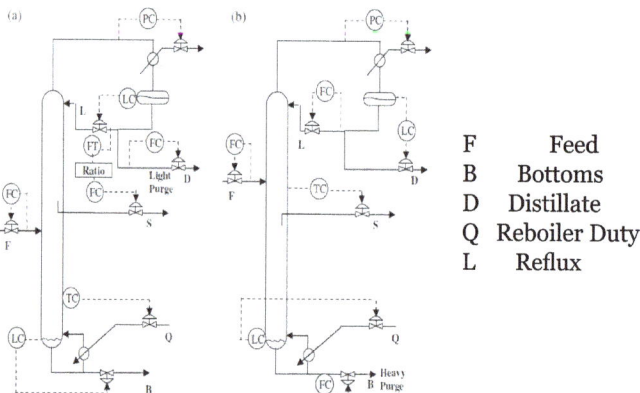

F	Feed
B	Bottoms
D	Distillate
Q	Reboiler Duty
L	Reflux

Control of purge columns of (a) Liquid side draw and (b)Vapor side draw

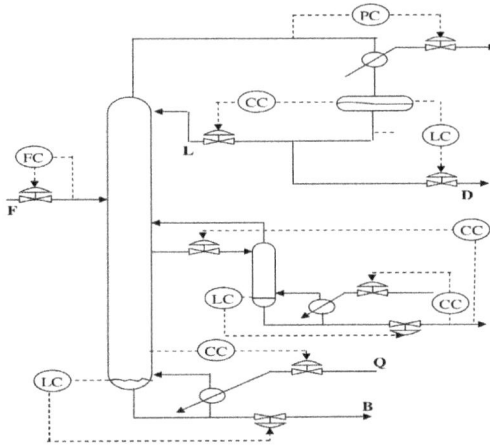

(a). Side stream column with stripper

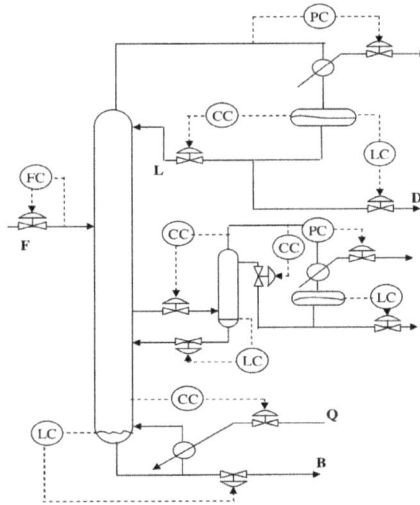

(b). Side stream column with rectifier

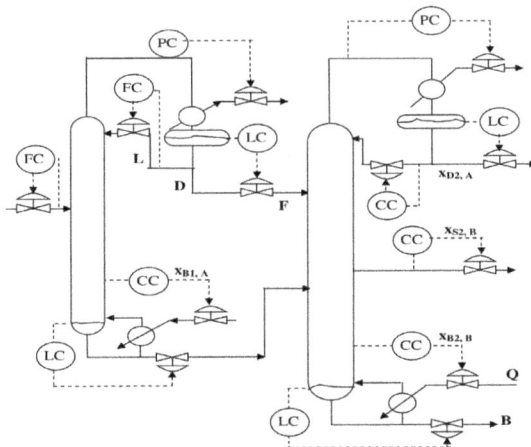

Side stream column with prefractionator

Control of Heat Integrated Columns

Heat integration arrangements in columns consist of the hot vapour from high pressure column providing the energy for reboil in a low pressure. The reboiler for the low pressure column then also acts as the condenser for the high pressure column. Three possible heat integration schemes are shown in Figure. In the feed split scheme, a binary fresh feed is split and fed to two columns. One of the columns is operated at high pressure and the other at low pressure. The pressure difference is chosen so that the hot vapour is 10-15 C hotter than the low pressure column reboiler temperature. The temperature difference provides the driving force for reboiling the liquid in the low pressure column. In the control structure shown, note that the feed to the low pressure column is adjusted so that the bottoms composition is maintained. This is because the reboiler duty in the low pressure column cannot be manipulated. Heat integration thus leads to the loss of a control degree of freedom. Also note that this heat integration scheme is used for a binary separation as the presence of LLK / HHK components can affect the column temperature profiles sufficiently so that the temperature driving force necessary for heat transfer in the low pressure column reboiler disappears.

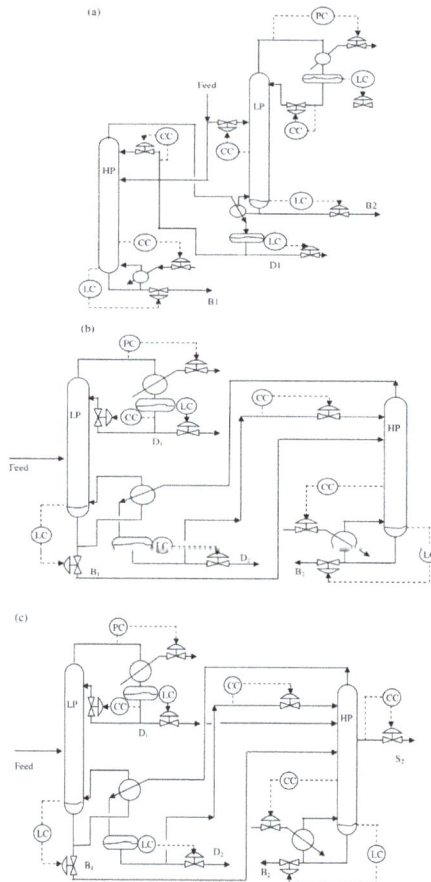

Heat integrated columns (a) Split of feed (binary); (b) Light split reverse (binary);
(c) Prefractionator reverse (ternary)

Figure (b) shows the reverse light split heat integration scheme. Approximately, half the light component is removed as the distillate from the first low pressure column. The bottoms is fed to the high pressure column to remove the light and heavy components as the distillate and bottoms respectively. The hot vapour from the top of this column is condensed to provide vapour reboil into the low pressure column. The direction of heat integration is reverse to that of the process flow. Hence the name, reverse light split. The forward light split configuration is also possible. The control structure for this heat integration scheme is self-explanatory.

Figure (c) shows another heat integration scheme that can be used for ternary mixtures. The scheme is the same as prefractionator side-draw complex column discussed previously except for reboiler in the pre-fractionator (low pressure column) acting as the condenser for the main column (high pressure) through heat integration. The control structure for this configuration is again self explanatory.

Control of Homogenous Extractive Distillation System

Homogenous extractive distillation is used to separate a binary AB mixture that cannot be separated due to relative volatility approaching one or the presence of a binary azeotrope. As shown in Figure, A heavy solvent S is added near the top of the first column, the extractive column, to soak in one of the components (say B). The distillate from the first column is then near pure A. The bottoms, a mixture of S and B, are fed to the solvent recovery column that recovers the heavy S in the bottoms and recycles it back to the extractive distillation column. The distillate from the solvent recovery column is pure B. The control scheme depicted in the Figure manipulates the reboiler duty in the two columns to keep respectively A and B from falling down the bottoms in the two columns. The solvent into the extractive column is ratioed to ensure fresh feed rate to ensure enough solvent for extracting component B. Note that the bottom sump level is not controlled and the sump must provide enough surge capacity for handling the expected variation in the fresh feed flow rate. Any loss of the solvent over long time is made up by a make-up solvent stream (not shown).

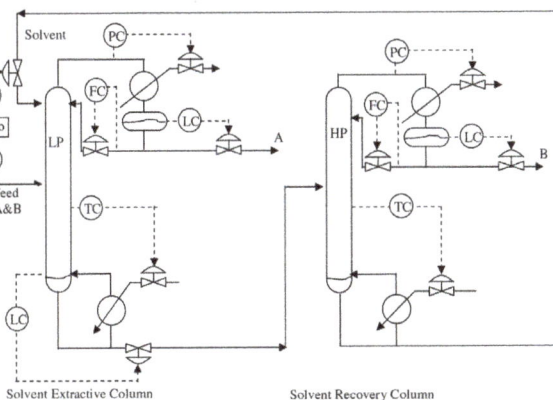

Control of Extractive distillation column

Plant-wide Considerations

In the plant-wide context, the distillate and / or bottoms would feed into a downstream unit such as another distillation column. The variation in the distillate / bottoms flow rate then acts as a disturbance into the downstream unit. The LQ control structure is particularly preferable as the reflux drum and reboiler levels are controlled using the P only controller which results in a smooth flow change into the downstream unit. If however, the DQ (or LB) control structure is used in a high reflux ratio (reboil ratio) column and a tray temperature is controlled using D, the reflux drum level controller manipulates L and must be tightly tuned for a fast dynamics of the of closed loop temperature controller. The D would then show large changes disturbing the downstream unit. Feed-forward control action can and should be used to mitigate the propagation of variability to downstream units. For example, the distillate rate may be ratioed to the feed rate with the composition / temperature controller setting the ratio set-point. Alternatively, the distillate may be moved in ratio with reflux with the composition controller setting the reflux ratio set-point. The reflux level controller is then tuned as a P only controller for smooth changes in the reflux and hence the distillate. The variability in the distillate rate can thus be greatly reduced improving the overall plant-wide control performance.

The vapour distillate from a partial condenser can be manipulated to control the tower pressure. This is however not a good idea if the vapour stream feeds directly into a downstream unit and not a surge tank. In such a scenario, the column pressure should be controlled using the condenser cooling duty, the reflux drum level controlled using the reflux rate and the vent rate moved in ratio with the reflux. This arrangement mitigates the propagation of variability downsteam.

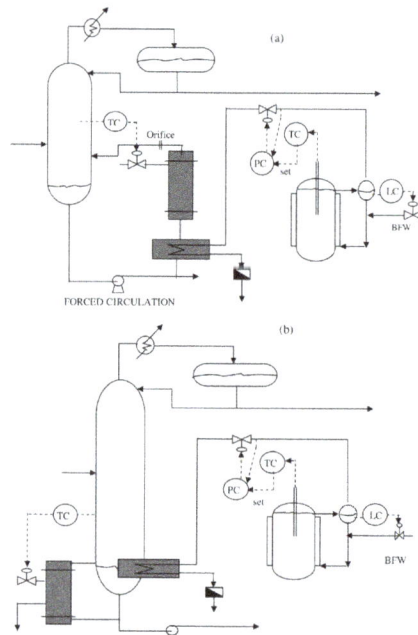

Reactor/column heat integration with auxiliary reboiler in (a) series (b) parallel

The control of energy integrated distillation columns can also be problematic as a disturbance on the hot vapour side necessarily affects the boil-up in the reboiler using the hot vapour as the heat source (instead of steam). To maintain the control tray temperature in the heat integrated column, an auxiliary reboiler (or condenser, as appropriate) is provided. The heat integrated reboiler and the auxiliary reboiler may be arranged in a parallel or a series arrangement. The series arrangement is preferred as the temperature variations in the hot vapour are attenuated due to variation in the temperature driving force in the auxiliary reboiler as shown in Figure (a). In the parallel arrangement, the auxiliary reboiler must adjust for the variability on the hot vapour side after it has entered the column. One way to prevent this is to use a total heat input controller as shown in Figure.

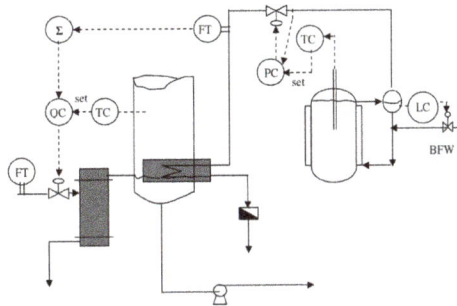

Reactor/column heat integration with auxiliary reboiler in parallel and Q controller

Control of Reactors

A reactor is the heart of a chemical process where the reactants undergo the desired chemical transformation to the products. The transformation is usually incomplete and is also accompanied by undesirable transformation to by-products through side reactions. The reactor operating conditions of temperature, concentration and flow rates determine the production rate of the main product and the side-products. The downstream separation load for separating the unreacted reactants from the products and recycling them back to the reactor is also determined by the reactor. Proper operation and control of the reactor is then crucial to determining the overall process operating profit. The reactor conversion (or yield) and selectivity are the two most commonly used reactor performance metrics that are directly related to the economics of the process. The conversion is defined as the fraction (or %age) of reactant in the feed that reacts to form product(s). The yield is the conversion of a key reactant, usually the limiting reactant. The selectivity is defined as the desired product generation rate relative to the total product generation rate (including all undesired side products). The yield and selectivity represent two key economic objectives of any process. A lower conversion would result in greater energy consumption to separate the unreacted reactants from the products and recycle them back to the reactor. The energy cost per kg product would then go up. A low selectivity represents an economic loss as more of the costly reactant gets converted to the undesired product with lower (or worse, a negative) profit margin.

Typically, as the conversion increases, the selectivity decreases so that the usual philosophy is to operate the reactor near the maximum conversion for which the selectivity is acceptable (say >95%).

Even as proper reactor operation is the key to profitability, controlling a reactor offers unique challenges as most reactions are accompanied by the generation or consumption of heat. The reaction heat generation / consumption alters the temperature of the reaction mixture which in turn affects the reaction rate and hence the rate of heat generation / consumption. The coupling of the thermal and reaction effects leads to a highly coupled non-linear system. The reaction rate, r (kmol.s^{-1}.m^{-3} or kmol.s^{-1}.kg^{-1} catalyst), for an irreversible reaction A + B → C would generally vary as

$$r = k.c_A^{\alpha}.c_B^{\beta}.$$

In the above expression, c_A and c_B are the concentrations (kmol.m^{-3}) of A and B respectively, k is the reaction rate constant, and α and β are the reaction order (typically > 0) with respect to A and B respectively. The units of the reaction rate constant depend on the reaction order and it follows the Arrhenious temperature dependence as

$$k = k_0.e^{(-E/RT)}$$

where E is the activation energy and k_0 is the Arrhenius frequency factor.

The form of the kinetic expressions above shows that the reactant concentration and the reaction temperature are the two basic manipulation handles for adjusting rate of product generation in a reactor. The reaction rate doubling for every 10 deg C increase in the temperature is an oft quoted rule of thumb. The reactor temperature is thus a dominant variable that significantly affects the reaction rate. When the reactant concentration is adjusted for changing the production rate, altering a key reactant concentration affects the reaction rate more than changing the concentration of other reactants. This is because most reactors must be operated such that one of the reactants is limiting, i.e. in excess of other reactants. The non-stoichiometric environment is necessary to suppress side reactions. For example, consider the main irreversible reaction A + B → C. The product C can further react irreversibly with B to form an undesired product D as C + B → D. For this reaction scheme, the reactor must be operated in an excess A environment so that the limited availability of B for further participation in the side-reaction suppresses by-product generation. Clearly, changing the limiting reactant B concentration would affect the reaction rate more than changing the excess reactant A concentration. In many industrial reactors, the reactor temperature and the limiting reactant concentration are the two dominant variables that are directly / indirectly adjusted for changing the product generation rate.

In exothermic reactions, the use of a selective catalyst lowers the activation energy for the main reaction. The activation energy for the side reactions is thus more than for the main reaction. In case the temperature is increased for increasing the production rate, the Arrhenius temperature dependence of the rate constant causes a larger relative change in the side- reaction rate. Thus if the main reaction rate increases by say 5%, the side reaction rate would increase by more than 5% (say ~10%). The reaction selectivity thus goes down. The adjustment of the reactor temperature for increasing the production rate thus must consider the detrimental effect on selectivity. In many industrial reactors, the reactor temperature is usually adjusted to compensate for catalyst poisoning / deactivation so that the overall reaction rates do not decrease over time.

Basic Reactor Types

The continuous stirred tank reactor, the plug flow reactor and the packed reactor are the most common reactor types used in the continuous process industry. These basic reactor types are shown in Figure. The PFR and PBR are similar except that the latter holds a catalyst bed to facilitate the reaction. The CSTR and the PFR (or PBR) differ fundamentally in terms of back mixing. In the PFR (and PBR), the fluid travels along a pipe as a plug so that every atom entering the reactor spends the same amount of time inside the reactor before exiting. This time is also referred to as the reactor residence time. Plug flow, by definition, implies no back mixing. The exact opposite of plug flow is perfect back mixing as in the CSTR. The back mixing is accomplished using agitators, spargers and fluidization.

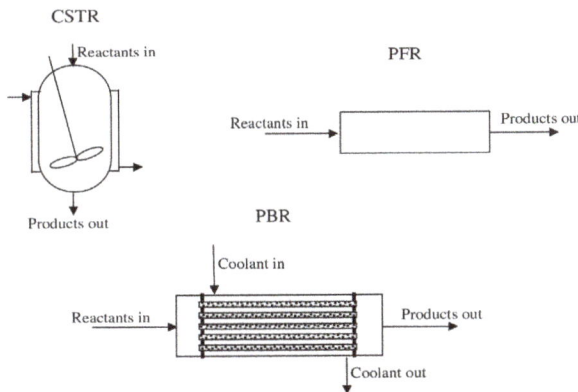

Basic reactor types

Plug Flow Reactor

PFR Basics

To understand the behaviour of a PFR (or PBR), imagine a plug of fluid flowing through the reactor. As it moves through, the concentration of the reactants goes down as they undergo reaction. Assuming adiabatic operation and an exothermic reaction, the heat released due to reaction would heat up the plug. The increase in the plug temperature causes the reaction rate to increase further so that the temperature in the initial part of the reactor increases exponentially. At a sufficient length down the reactor, the reaction rate begins to decrease due to the limited availability of reactants. For a large enough reactor length, the reaction rate would go to zero as the limiting reactant gets exhausted. The temperature profile for an adiabatic PFR thus resembles a sigmoid as shown in Figure. The difference in the inlet and outlet temperature is referred to as the adiabatic temperature rise. If the reaction is highly exothermic, the adiabatic temperature rise is large, which is usually unacceptable due to reasons such as promotion of side reactions at the higher temperatures, possibility of catalyst sintering in a PBR, increase in the material of construction cost etc. The cooled PFR / PBR is then used.

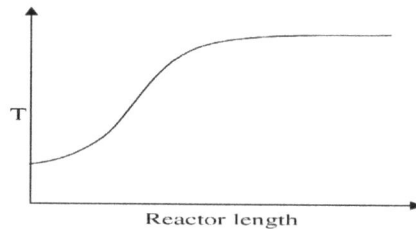

Temperature profile for an adiabatic PFR

The most common cooled PBR arrangement is shown in Figure. Catalyst is loaded in the tubes of a shell and tube heat exchanger. Pressurized water is re-circulated on the shell side. The water carries the reaction heat to form steam in the steam drum. The high re-circulation rate ensures a near constant temperature on the shell side. Unlike adiabatic operation, the temperature profile initially increases as the rate of heat generation is more than the cooling rate and later decreases with the reaction rate decreasing due to reactant depletion and the cooling rate increasing due to the higher temperature driving force. The temperature profile thus exhibits a maximum, also referred to as the hot spot. For highly exothermic systems, the reactor temperature profile can be extremely sensitive to the operating conditions, in particular the reactor inlet temperature and the shell side temperature.

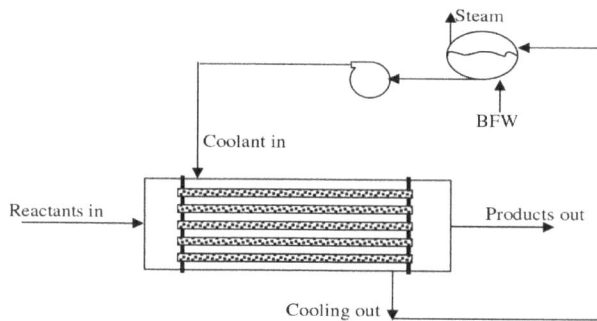

Cooled Packed bed reactor

If the reactant inlet temperature (or coolant temperature) is too low, reaction may not kick in leading to the quenched state. If the reactant inlet temperature is too high, the reaction can proceed so fast that only a small fraction of the heat released gets removed resulting in a temperature run-away. The quenched, hot-spot and run-away reactor temperature profiles are illustrated in Figure.

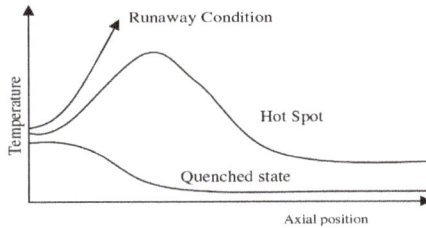

Cooled plug flow reactor temperature profiles

The adiabatic or cooled tubular reactor is commonly used in the process industry for gas phase reactions. The reactor is usually operated at the maximum equipment pressure so that the reactant partial pressures are as high as possible for maximum reaction rate. This also reduces the design volume of the reactor for a given conversion. The control of PFRs is considered next.

Control of PFRs

Adiabatic PFR

The adiabatic PFR is the simplest reactor configuration and is used when the adiabatic temperature rise is acceptable. The reactants (fresh + recycled) are heated to the reaction temperature using a furnace. The furnace heat duty holds the reactor inlet temperature constant. This is shown in Figure (a). The adiabatic temperature rise sets the reactor outlet temperature. Sometimes the outlet temperature is also controlled to maintain the reactor conversion and selectivity. The limiting reactant fresh feed rate may be used as the manipulation handle as illustrated in Figure (b).

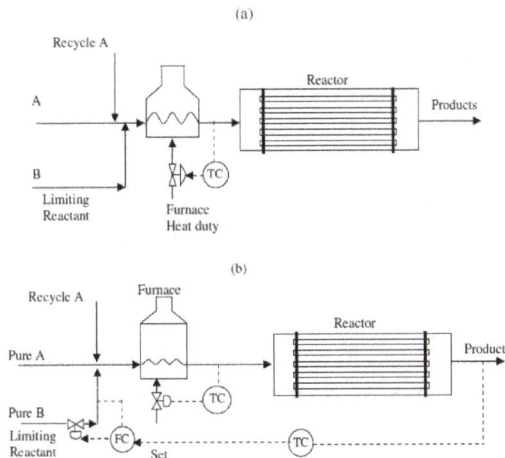

Simple adiabatic PFR structures

Cooled Tubular Reactors

Cooled tubular reactors are more challenging from the control perspective. The hot spot temperature must be tightly controlled to prevent a runaway. This is accomplished using auctioneering temperature control as illustrated in Figure. The measurements from an array of thermocouples (or RTDs) placed along the length of the reactor are input to a high selector that passes the maximum temperature to the hot spot temperature controller. This controller typically manipulates the reactor cooling duty. The control structure for the most common cooled PBR arrangement is shown in Figure. The temperature controller sets the steam drum pressure set-point. A change in the drum pressure alters pressurized water boiling point which in turn changes the temperature driving force for heat removal from reactor.

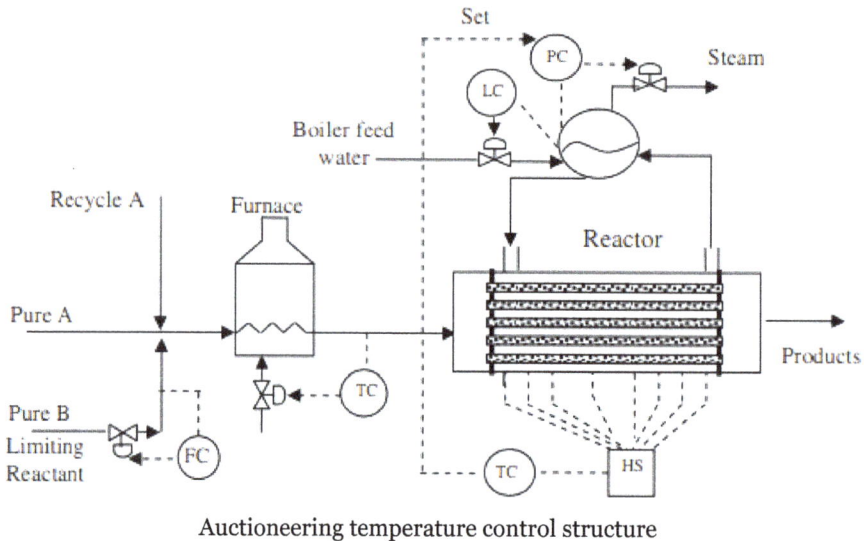

Auctioneering temperature control structure

In addition to the reactor cooling duty, there are two other possible manipulation handles for reactor heat management, namely the reactor inlet temperature set-point and the limiting reactant flow rate into the reactor. The schemes are shown in Figure. Both the schemes work by changing the heat generation due to reaction. The non-linearity between the controlled and manipulated variable is severe in all the three schemes. This may be understood using the analogy of a fire. It is very easy to make a fire whereas extinguishing one requires much effort. Similarly, it requires much more control effort to adjust for an increase in the hot-spot temperature than for the same decrease. The controller thus must be aggressive for deviations in one direction and not too aggressive in the other. Gain scheduling is sometimes used with the magnitude of the controller gain depending on the magnitude and sign of the error signal (or a more sophisticated schedule). The possibility of temperature runaway also requires that large overshoots above the set-point be avoided even as the controller is aggressive. The derivative action is often employed to suppress closed loop oscillations.

(a)

(b)

Two other possible manipulation handles for reactor heat management

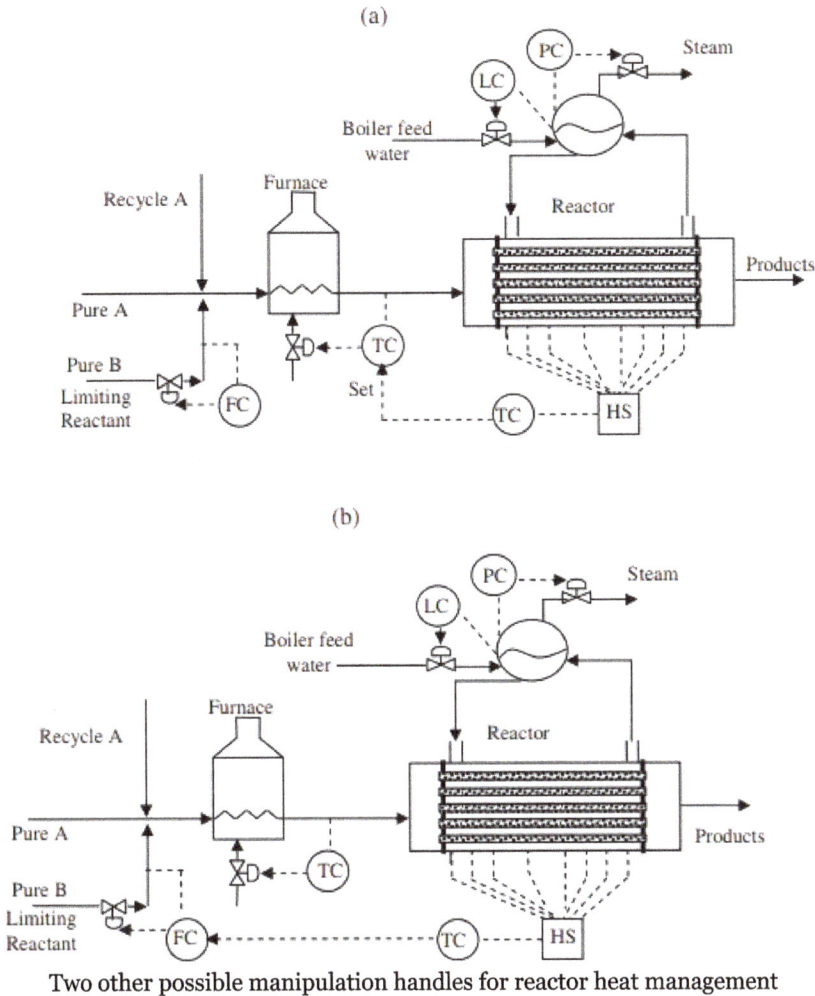

It is noted that controlling the reactor outlet temperature may sometimes provide adequate regulation of the hot spot temperature. The applicability of this much simpler scheme depends on how close the outlet is to the reactor hot spot at the base design condition. When the hot spot is close to the reactor outlet, a change in the outlet temperature correlates well with the change in the hot spot temperature so that controlling the outlet temperature provides adequate regulation of the hot spot temperature.

Intermediate Cooling and Cold–Shot Cooled Reactors

Two other commonly used heat removal configurations, namely, intermediate cooling and cold shot cooling, are shown in Figure along with the control structure. Explicit intermediate coolers are provided in equilibrium limited exothermic reactions to increase the overall equilibrium conversion. Cold shot cooling is frequently employed in polymerization reactors where it is extremely important to hold the temperature profile in the reactor to maintain the molecular weight and polydispersity of the product polymer.

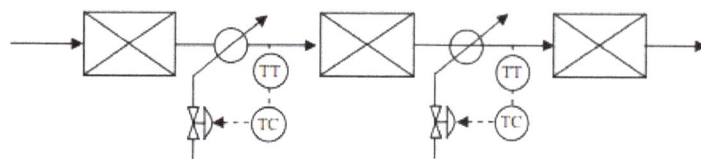

(a)

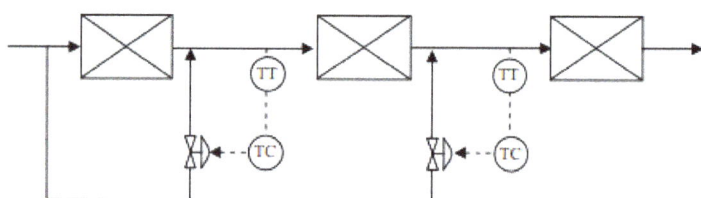

(b)

(a) Intermediate cooling in sequence reactors
(b) Cold shot cooling

Modern chemical plants frequently recover the reaction heat using heat integration. A common heat integration scheme employs the feed effluent heat exchanger to preheat the cold reactor feed using the hot reactor effluent gases as in Figure. The heat integration results in energy recycle loop which can lead to instability. Consider the extreme case of a FEHE that heats the reactants to the reaction temperature so that there is no furnace. If the temperature of the cold reactants rises, the reactor inlet temperature will increase. This would cause more reaction accompanied by heat release resulting in an increase in the hot reactor effluent gas. The hotter effluent would cause a further increase in the reactor inlet temperature resulting in a temperature runaway or instability. This instability can be prevented if the reactor inlet temperature (or outlet temperature) is controlled. The furnace performs this function by breaking the positive thermal feedback loop. An alternative to the FEHE is to recover the reaction heat as steam in a waste heat boiler and feed it to the steam utility network. This removes the thermal feedback while ensuring almost 100% heat recovery.

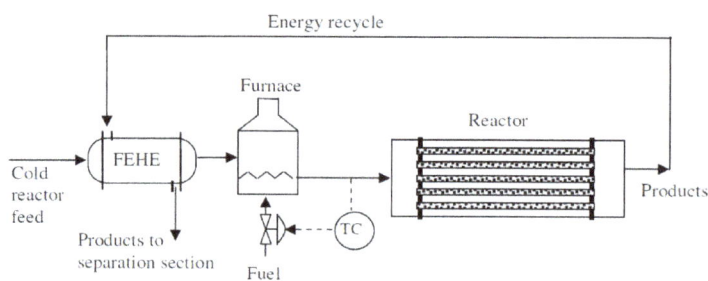

Heat integration scheme

Catalytic packed bed reactors differ from simple plug flow reactors in that that the catalyst bed constitutes a significant thermal capacitance. The heat capacity of the packed bed can sometimes lead to the outlet temperature exhibiting an inverse response with respect to the inlet temperature. This effect is due to the difference in the propagation rate of the gas and the bed thermal effect through the reactor. If the reactor inlet temperature decreases, the bed temperature does not decrease immediately. The cooler gas plug thus comes in contact with the hot packing and heats up. The reaction now kicks off leading to the outlet temperature increasing. Once the catalyst bed cools down, the outlet temperature of course decreases. This inverse response or "wrong" way behaviour destabilizes the control loop so that a PID controller must be detuned. In cases where the closed loop performance is not satisfactory, the application of advanced control techniques such as the Smith predictor is recommended.

Continuous Stirred Tank Reactor

Perfect back-mixing occurs in ideal continuous stirred tank reactors. Due to the mixing, the composition at the reactor is the same as the composition inside the reactor. The reaction thus occurs at the reactor outlet concentration. The reactant, upon entering the reactor, thus gets diluted instantaneously to the lower reactor composition. Since the reaction occurs at the exit conditions, the steady state conversion now depends on both the inlet and outlet conditions. This creates higher non-linearity so that the existence of multiple steady states is a distinct possibility. This is in direct contrast to PFRs where the outlet conditions are uniquely determined by the inlet conditions. Back-mixing thus causes material feedback with the possibility of multiple solutions.

Consider a jacketed CSTR as in Figure. Assume that the coolant flow rate is high so that the jacket temperature is nearly constant. The heat removal rate varies linearly with the jacket-reactor temperature difference. The heat generation due to reaction is an S-shaped sigmoid with respect to the reactor temperature. At steady state the heat removal rate and the heat generation rate must balance each other. The reactor can exhibit a unique steady state or multiple steady states. These scenarios are depicted in Figure. For multiple steady states, there are three steady states corresponding to a high, intermediate and low temperature. The high and low temperature steady states are stable while the intermediate temperature steady state is unstable. This is because around the intermediate steady state, if the temperature increases slightly, the rate of heat generation increases more rapidly than the rate of heat removal so that the temperature would continue to rise and not return back to the intermediate steady state ie an open loop unstable system. In contrast, at the high / low temperature steady state, the slope of the heat removal curve is more than the heat generation curve in the vicinity of the steady state implying stable open loop behaviour. Typically, reactor operation at the intermediate steady state is desired as the high temperature steady state may lead to catalyst sintering and undesirable side reactions while the reaction rate is small at the low temperature steady state. For such open loop unstable reactors, a temperature controlled must be used to stabilize the reactor. The closed loop system becomes stable for a controller gain above a critical value. At lower gains, the feedback action is not enough to stabilize the unstable system. For extremely large controller gains, the closed loop system becomes unstable due to too much feedback, similar to open loop stable processes. The closed loop system is thus stable only for a range of controller gain. Conventional tuning rules such as the ZN / TLC procedure should therefore not be applied. Multiple steady states are avoided when a large heat transfer area is provided. Unique solution CSTRs are much easier to control and the heat transfer system for CSTRs should be properly designed to avoid multiplicity.

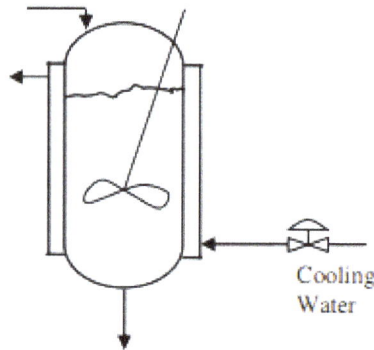

High flow rate cooling water temperature control

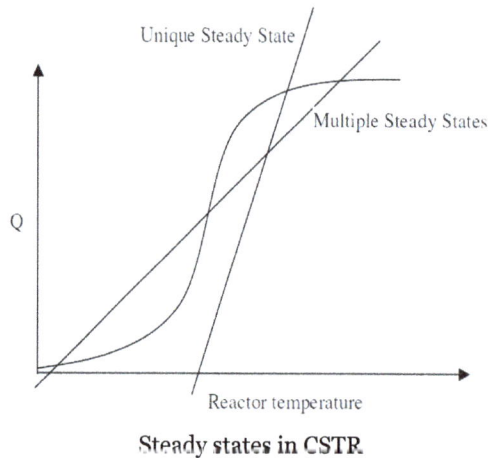

Steady states in CSTR

Jacket Cooled CSTR

Evidently, proper regulation of the reaction heat removal is one of the main control tasks. The simplest heat transfer arrangement is that of a jacketed CSTR with cooling water flowing along the jacket. Its flow rate is adjusted to maintain the reactor temperature as shown in Figure. This approach suffers from drawbacks such as a non-linear process gain due to variation in the heat transfer coefficient with cooling water flow and local reactor hot spots due to the jacket temperature profile. These problems can be mitigated by using a recirculation loop as shown in Figure (a). The recirculation allows for a constant coolant flow rate so that the jacket temperature is constant. The reactor temperature is maintained by adjusting the fresh coolant flow rate into the recirculation. The recirculation loop introduces and additional thermal lag into the heat transfer loop resulting in a slow closed loop response. The use of a cascade control scheme as in Figure (b), where the jacket temperature is tightly controlled by adjusting the coolant flow (slave loop) and the reactor temperature loop adjusts this set-point, significantly improves the closed loop reactor temperature control.

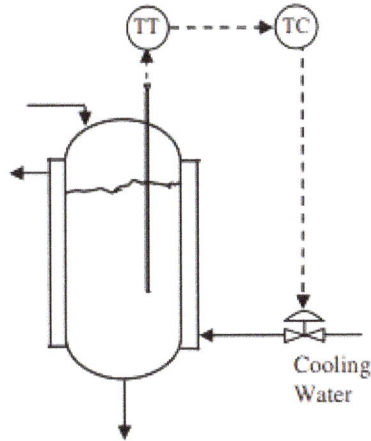

Simplest CSTR Control Structure

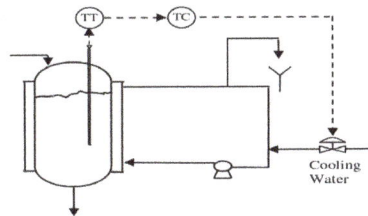

(a)

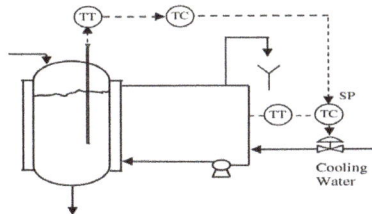

(b)

(a) Circulating cooling water temperature control
(b) Cascade control

Reaction Heat Removal as Steam

A common arrangement for recovering the reaction heat is to generate steam from the hot pressurized water recirculating in the jacket recirculation loop. The heat removal scheme and the control structure are shown in Figure. The level in the steam drum would exhibit the inverse response so that the boiler feed water flow into the drum is ratioed to the steam flow with the level controlled adjusting its set-point. This arrangement allows for the feed water to move in the correct direction for load changes.

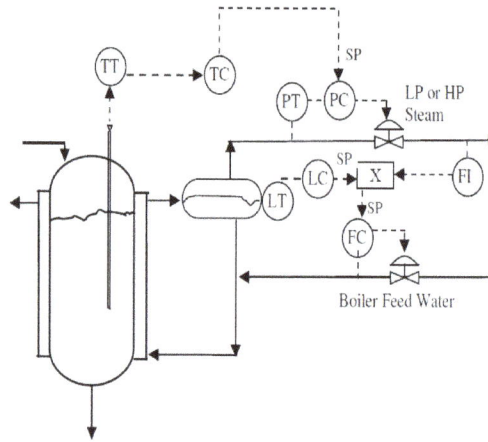

Temperature control through steam generation

External Heat Exchanger

In high temperature applications, the generation of steam is not possible so that expensive proprietary coolant oils must be used. An external heat exchanger is then provided as in Figure for a closed circuit coolant loop. Cooling water is used as the coolant in the external heat exchanger to remove the reaction heat carried by the hot oil. The control structure shown in the Figure is self explanatory.

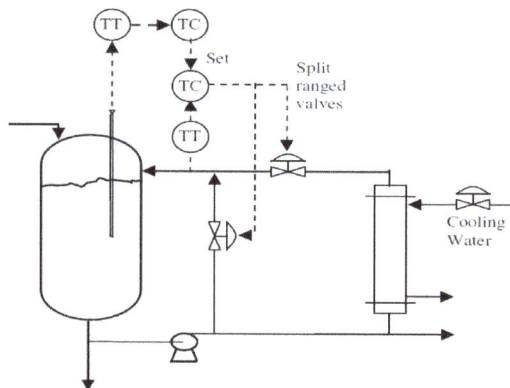

Use of external Heat exchanger for temperature control

Cooling Coils

In the jacketed CSTR, the heat transfer area is determined by the reactor volume and may not be sufficient. Cooling coils, as in Figure, are used for higher heat transfer area per unit volume. The control scheme adjusts the coolant flow rate for maintaining the jacket temperature.

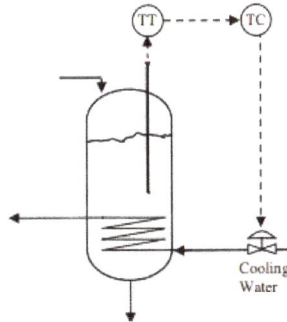

Extended heat transfer area through use of cooling coils

External Cooling by Content Recirculation

Another alternative for removing the reaction heat when the jacket heat transfer area is insufficient is to circulate the reaction mixture through an external heat exchanger and feed it back into the reactor. This scheme is shown in Figure. The reactor temperature is maintained by manipulating the temperature of the cooled reaction mixture from the external heat exchanger. The cooling water flow rate into the heat exchanger is adjusted to maintain the cooled reaction mixture temperature. In this control scheme, the external heat exchanger introduces a significant thermal lag into the slave loop. The dynamics of the slave loop can be significantly improved by a slight design modification providing for bypassing a small fraction of the reaction mixture stream around the external heat exchanger. This is illustrated in Figure. The thermal lag is thus replaced by the negligible mixing lag as the dominant time constant of the slave loop resulting in significantly improved reactor temperature control.

Boiling CSTR with External Condenser

When the reaction mixture boils, excellent reactor temperature control can be achieved using an external condenser that condenses the vapour and refluxes the cold condensate back into the reactor. The arrangement is shown in Figure. Note that the condensate flows back into the reactor by gravity so that the condenser should be at a sufficient elevation above the reactor. The U-leg seal is provided to force the vapour to enter the condenser from the correct entry port. Note that the reactor temperature in this case is self-regulatory.

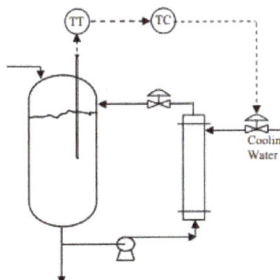

Circulation of reactor contents through external heat exchanger

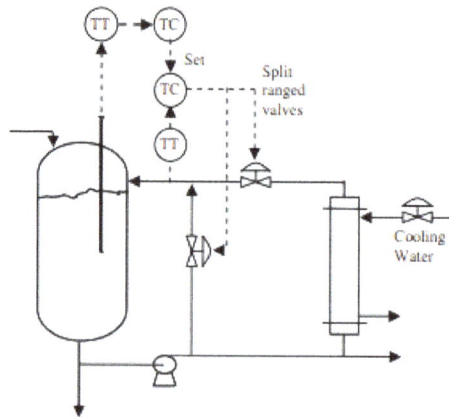

Bypass of circulating reactor content around external Heat exchanger

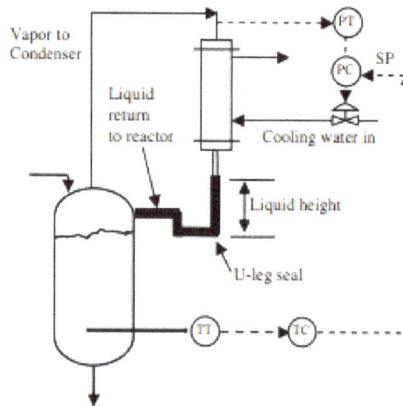

Cooling through vaporization of reactor content

Reactor Heat Removal Capacity Constraint

A possibility for controlling the reactor temperature when the heat transfer area is limiting is to adjust the reactant feed rate so that the heat generation due to reaction changes appropriately. The cooling water valve is fully open. The scheme is shown in Figure. While, appealing in its own right, this control strategy should not be implemented in practice (or used with due caution) as the open loop dynamics of the temperature loop is slow due to the composition lag introduced by the reaction mixture volume. As the feed rate changes, the composition of the reaction mixture changes slowly due to the large reactor hold up. The reaction heat generation thus changes slowly. In case the reactor temperature goes down, the temperature controller would add more feed. The unreacted reactant amount in the reactor thus goes up. Once the reactor temperature begins to increase, reaction would "kick-in" due to the large amount of unreacted reactant inside the reactor. The possibility of a reactor run- away is thus always lurking in the back-drop, especially for highly exothermic reactions. The scheme may be workable for mildly exothermic reactions.

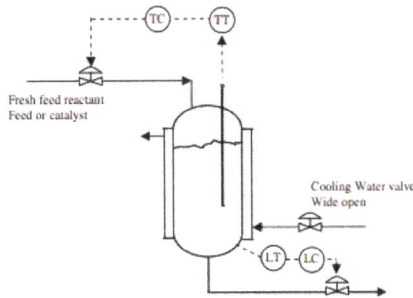

Temperature control by manipulating the fresh feed rate

The problem of reactor runaway can be circumvented by the use of a valve positioning scheme as illustrated in Figure. The reactor temperature is controlled by adjusting the reactor cooling duty. The valve positioning controller measures the cooling duty valve position and slowly adjusts the feed rate so that eventually the cooling duty valve is near fully open and the reactor operates at maximum through-put. In this scheme, the temperature control loop effectively rejects short term disturbances as its dynamics are much faster compared to controlling the reactor temperature directly using the fresh feed rate. Over the long term, the VPC ensures the reactor is operating at near maximum cooling duty, ie maximum through-put. Note that the feed rate can be directly manipualted in a PFR since the material flows through as a plug and there is no back-mixing implying little / no build-up of unreacted reactants inside the reactor.

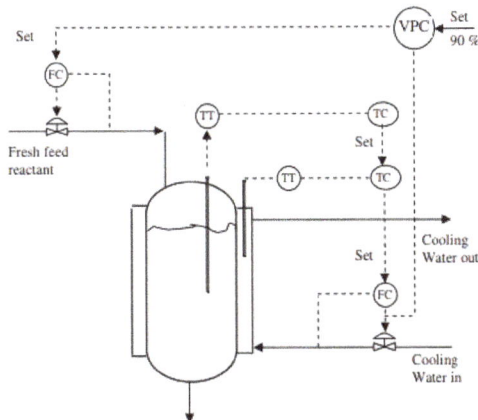

Valve positioning control for throughput maximization

Heat Exchanger Control

Heat exchangers are widely used for heating / cooling process streams to the desired temperature or to change the phase of a stream. The heat exchanger is thus for removing / adding sensible or latent heat. Figure shows the schematic of a counter-current shell and tube heat exchanger. The hot stream flows through the tubes and loses its heat to the cold stram flowing through the shell. The heat exchange is driven by the temperature difference between the shell side and the tube side. For a given inlet temperature of the hot and cold streams, the temperature driving force is more for the counter-current flow arrangement. Most exchangers are thus operated with counter-current flow.

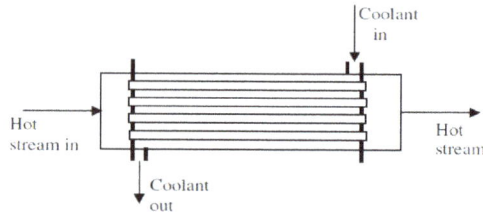

Counter current Shell and Tube heat exchanger

The heat exchangers in a process can be usefully classified into utility heat exchangers or process-to-process heat exchangers. Utility heat exchangers typically use steam or cooling water to respectively add or remove heat from a process stream. In process-to-process heat exchangers, both the hold and cold streams are process streams.

Control of Utility Heat Exchangers

The purpose of a utility heat exchanger is to provide (remove) as much heat as is necessary to maintain the process stream temperature. A simple temperature controller that adjusts the utility flow rate to maintain the process stream temperature accomplishes this function. Figure (a) shows a cooler that uses cooling water for heat removal. Figure (b) shows a heater using steam as the utility fluid. The control loops are self-explanatory.

Sometimes the heat transfer is controlled without adjusting the utility flow. For example in a partial condenser that vents the non-condensables as a vent stream, the cooling water valve is fully open and the vent rate is adjusted to control the condenser pressure. The pressure sets the dew point temperature of the condensables in the vapour stream fixing the temperature driving force across the tubes. The control scheme is illustrated in Figure.

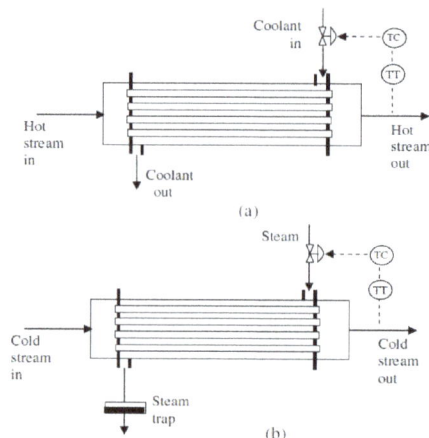

Control of utility exchangers

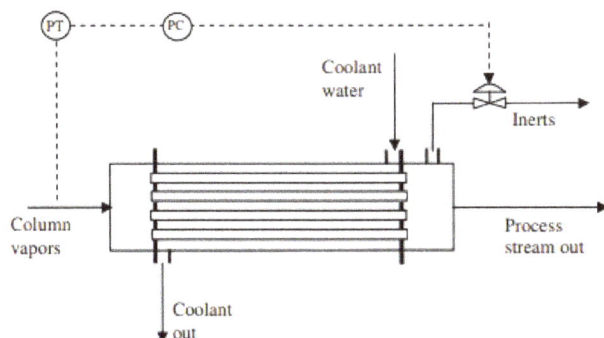

Indirect control of Heat exchanger using partial pressure of non-condensable

The flooded condenser is another common arrangement where the level of the condensate determines the number of tubes that are submerged. Heat is thus transferred only across the tubes exposed to the vapour. The cooling rate thus gets adjusted to maintain the pressure by manipulating the condensate draw which affects the level. The liquid hold up inside the condenser represents a significant lag (~2-5 minutes) so that the pressure cannot be controlled very tightly. Flooded condenser arrangement is shown in figure.

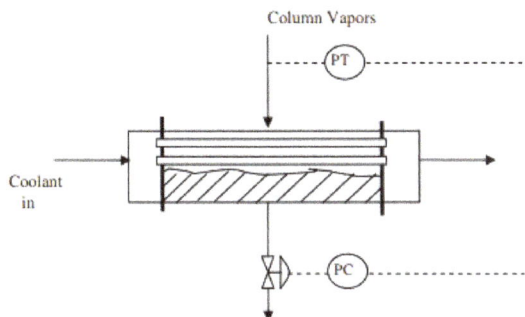

Heat transfer control by variable heat transfer area in flooded condenser

Control of Process-to-Process Heat Exchangers

Process to process heat exchangers transfer heat between two process streams. The flow of these process streams is usually set elsewhere in the plant so that adjusting the flow rate of one of the process streams to regulate the amount of heat transferred is not possible. To provide a control degree-of-freedom for regulating the heat transferred, a small by-pass (~5-10%) of one of the process streams around the heat exchanger is provided. The outlet temperature of this process stream or the other process stream can be controlled by manipulating the by-pass rate. These two schemes are illustrated in Figure. In the former, tight temperature control is possible as the amount of heat transferred is governed by the bypass. In the latter, a thermal lag of the order of 0.5 to 2 minutes exists between the manipulated and controlled variable.

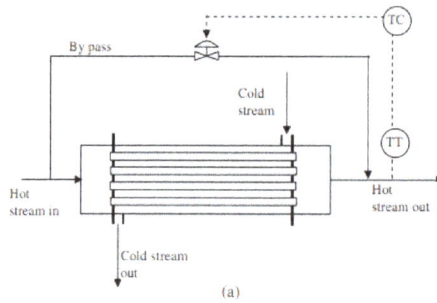

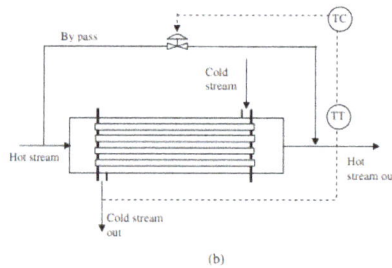

By-pass control of process to process heat exchangers
(a) Controlling and bypassing hot stream
(b) Controlling cold stream and bypassing hot stream

Process-to-process heat exchangers are increasingly used for heat integration to minimize the energy consumed per kg product. Given that the flow of the two process streams into the heat exchangers is set elsewhere, an auxiliary utility heat exchanger is often provided to control the temperature of the more important process stream. The size of the auxiliary utility heat exchanger should be large enough for effective disturbance rejection.

In the plant-wide context, heat integration using process-to-process heat exchangers causes interaction between the interconnected units. In particular, it is necessary to ensure that in the quest for maximum energy recovery, energy recycle circuits do not lead to instability. Sufficient control degrees-of-freedom should be provided in the form of auxiliary utility exchangers so that the variability is transferred to the plant utility system.

Control of Miscellaneous Systems

In this chapter, the control of other common units in the industry such as furnaces, compressors, refrigeration systems and plant utility systems is briefly described.

Furnace Controls

A furnace heats a process stream to high temperature using combustion of a fuel as the heat source. It consists of a fire box or combustion zone with tube bundles carrying the process stream to be heated. Fuel is burnt with air in the combustion zone to heat the tubes to very high temperatures. Typically a convective heat transfer zone is also provided in furnaces to recover heat from the hot flue gases. The furnace is essentially a reactor combusting fuel with air. The control objective is to satisfy the 'on-demand' heat load. The control system shown in Figure is typically used. The fuel-to-air ratio must be nearly stoichiometric for complete combustion of the fuel. Excess air is not fed in as that would increase the flue gas discharge rate. Less than stoichiometric air would lead to partial combustion or worse, unburnt fuel remaining in furnace. The flue gas oxygen concentration is a good indicator of the quality of combustion and adjusts the fuel-to-air ratio setpoint. The air is fed in ratio to the fuel. The forced draft fan speed is varied to change the air feed rate. An induced draft fan is provided at the outlet to suck the flue gases out of the furnace. Its speed is controlled to maintain the pressure inside the combustion chamber.

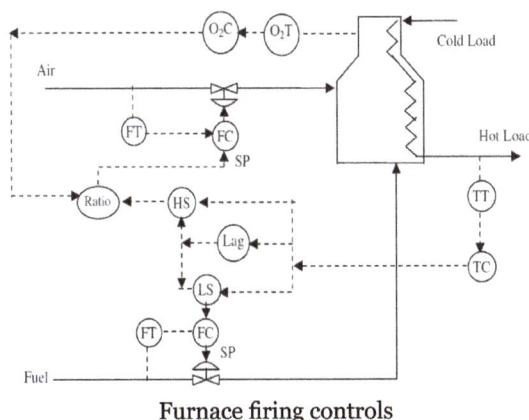

Furnace firing controls

A critical safety requirement is to operate the furnace such that the air is fed in excess during transients (load changes). This is necessary to ensure that all the fuel fed into the furnace is burnt and no unreacted fuel remains inside, lest it combust later to damage the furnace. Thus if the heat load increases, the air rate must be increased before the fuel valve is opened. On the other hand, if the heat load decreases, the fuel valve must be closed before the air flow is reduced. This control action is accomplished by lagging the heat load signal as shown in Figure. The lagged and the unlagged signals are then input to a high selector and a low selector. The output of the high selector sets the air flow controller set-point. The output of the low selector sets the fuel flow controller set-point. If the heat load increases, the high selector sends the unlagged signal to the air flow controller causing an instantaneous increase in the air flow. The low selector sends the lagged signal to the fuel flow controller. The fuel flow thus lags behind the air flow for an increase in the heat demand. For a decrease in the heat load, the high selector sends the lagged heat load to the air flow controller while the low selector sends the unlagged heat load to the fuel flow controller. The air thus lags behind the fuel for a heat load decrease. Furnace operation in excess air is thus ensured during transients.

Compressor Controls

Compressors are used to increase the pressure of gas stream. A cooler with a knock- out pot is typically provided at the compressor outlet to cool the hot pressurized gas and remove any condensables that liquefy due to the higher pressure and cooling. There are three important types of compressors used in plants, namely, centrifugal, axial and reciprocating. In reciprocating compressors, the through-put is adjusted by manipulating the strokes per minute or the length of a stroke. A recycle is always provided around the outlet of the compressor for the safety of the compressor.

Centrifugal compressors are similar to centrifugal pumps in that a rotating motor is used to impart energy to the fluid. To control the through-put, three configurations are typically used, namely, exit recycle, suction throttling and motor speed manipulation. These three schemes are illustrated in Figure. In the exit recycle scheme, a recycle around the compressor back to the inlet is provided which is adjusted to manipulate the through-put. Note that the total (recycle + fresh) flow rate through the compressor remains the same so that compressor operates at a single point on its characteristic curve. This is the most energy inefficient method of compressor operation. Also, note from the figure that the recycle is provided after the cooler so that energy recycle is prevented. In suction throttling, a valve is provided at the compressor suction and the through-put is manipulated by adjusting the suction pressure. At lower through-puts lesser energy is consumed as the amount of material flowing through the compressor is less. The most energy efficient method of throughput manipulation is to vary the rpm of a variable speed drive. High pressure steam, as in a turbine, is used many a times to provide the motive force for rotation. A cascade speed controller that adjusts the steam flow rate set-point maintains the drive speed. The drive speed set-point is input remotely by the through-put controller.

Another important consideration in compressor control is the prevention of surge at low flow rates. The compressor characteristic curve shows a maximum and the compression ratio dips a low flow rates due to compressibility. So much so, that if the flow rate goes low enough, the flow through the compressor can reverse direction. This causes the suction pressure to build and the flow almost immediately reverses direction again (i.e. flows out the compressor). This flow reversal cycle repeats in less than a second. To prevent the compressor surge phenomenon, the compressor discharge is recycled to the compressor suction. An anti-surge controller, as in Figure adjusts the recycle rate to prevent the flow through the compressor from dropping below a minimum. Note that this minimum must be sufficiently above the surge flow rate for the particular compressor rpm (or maximum rpm for variable speed drives).

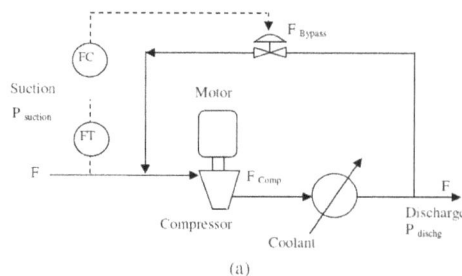

(a)

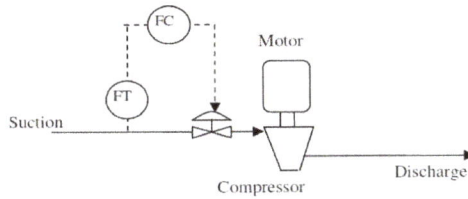

(b)

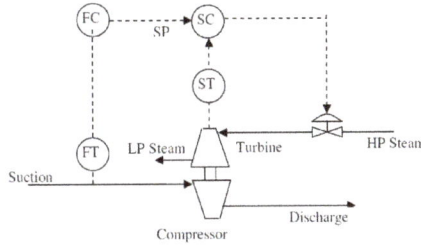

(c)

Compressor controls
(a) Exit recycle
(b) Suction throttling
(c) Motor speed manipulation

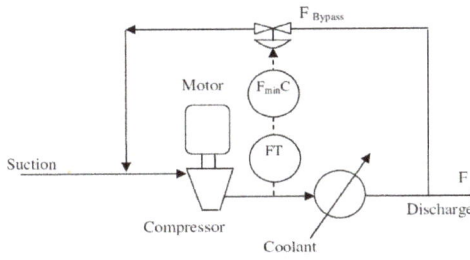

Compressor antisurge controller

Decanter Control

Decanters separate a heterogenous liquid-liquid mixture into its constituent liquid phases by utilizing the density difference between the liquid phases. The heterogenous mixture phase separates into a heavy and light liquid phase. Typically, the heavy liquid phase is aqueous while the light liquid phase is organic. Appropriately located withdrawal ports are provided in a decanter for removing aqueous and organic streams. To prevent the aqueous liquid from entering the organic liquid withdrawal port, the level of the liquid-liquid interface must be controlled. Also the organic phase level must be controlled. The simplest scheme, shown in Figure (a) manipulates the organic and aqueous stream flow rates to adjust the respective levels. The organic level controller must however interacts with the aqueous level controller. A simple and effective strategy for removing the interaction is to adjust the total flow out from the decanter to control the organic phase level. Figure (b) shows the corresponding control scheme. The organic stream flow is manipulated to maintain total flow (organic + aqueous) out of the decanter. The organic level controller sets the total flow set-point.

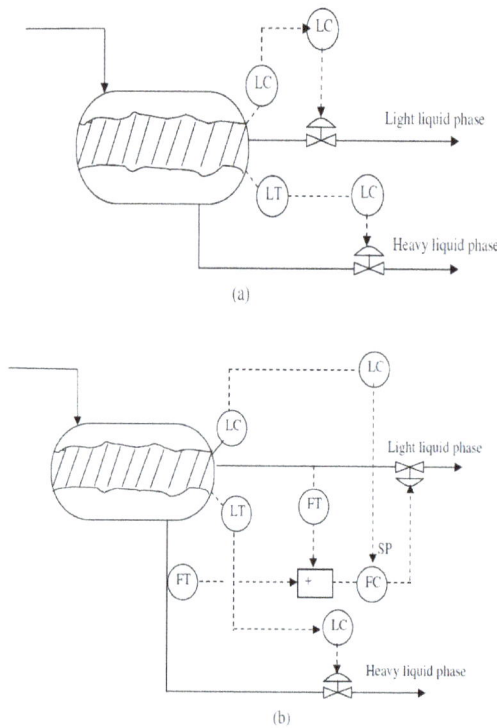

(a)

(b)

Decanter Controls
(a) Conventional level control
(b) Buckley control structure to eliminate interaction

Control of Refrigeration Systems

We study control schemes for the commonly used vapor compression and vapour absorption refrigeration cycle.

Vapor Compression Cycle

The refrigeration cycle typically employs compression. The cold refrigerant absorbs heat from the process stream and vaporizes in the evaporator. The vapour is compressed so that at the higher pressure, cooling water can be used to condense the refrigerant. The condensed refrigerant is collected in a surge drum and fed to evaporator. Figure shows control schemes for the compression refrigeration cycle. The chilled process stream temperature controller sets the evaporator pressure set-point. The evaporator pressure is controlled by adjusting the compressor suction valve. The level in the evaporator is controlled by adjusting the liquid refrigerant inlet valve. In case a variable speed drive compressor is used, the pressure controller is done away with and the temperature controller directly sets the drive speed set-point. The pressure controller is necessitated in the compressor suction throttling scheme to compensate for the throttling valve non-linearity. In the variable drive speed compressor, the variation in the suction pressure (evaporator pressure) with respect to the drive speed is relatively linear so the drive speed can be directly adjusted by the temperature controller. The level in the refrigerant surge drum is not controlled as the refrigerant forms a closed circuit. Notice that the heat transfer rate changes as the temperature driving force across the condenser changes due to the dependence of refrigerant boiling temperature on the evaporator pressure.

Vapor Absorption Cycle

In addition to compression systems, refrigerant absorption systems are also applied industrially. The absorption based refrigeration cycle and its control scheme is shown in Figure. Ammonia (refrigerant) rich strong liquor is distilled at high pressure to recover liquid ammonia as the distillate and ammonia lean weak liquor as the bottoms. The liquid ammonia is fed to the evaporator where it absorbs heat from the process stream to be chilled and evaporates. Vapor ammonia is absorbed by the 'weak liquor' water stream. The 'strong liquor' so formed is fed to the distillation column to completed the closed circuit refrigerant loop. The temperature of the chilled process stream is controlled by adjusting the level set- point of the evaporator. The heat transfer rate is thus varied by changing the area across which heat transfer occurs. The evaporator level controller adjusts the distillate liquid ammonia flow. An increase in the level of the evaporator implies an increase in the ammonia evaporation rate so that the weak liquor rate is increased in ratio to absorb the ammonia vapours. The strong liquor is cooled and collected in a surge drum. The level of the surge drum is not controlled. Liquid from the surge drum is pumped back to the distillation column through a process-to-process heater that recovers heat from the hot 'weak liquor' bottoms from the distillation column. The flow rate of the strong liquor to the column is adjusted to maintain the column bottoms level. Also, the steam to the reboiler is manipulated to maintain a tray temperature.

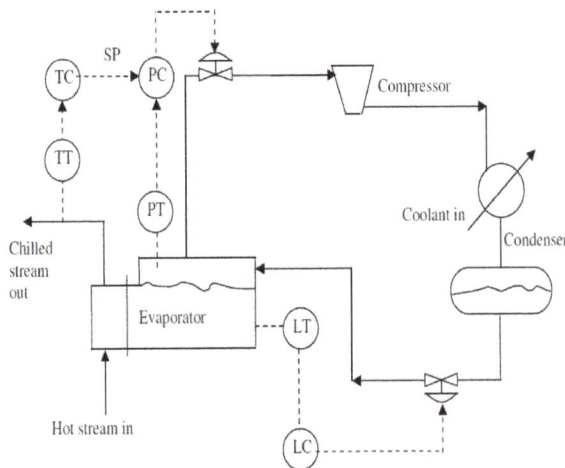

Compression refrigeration controls

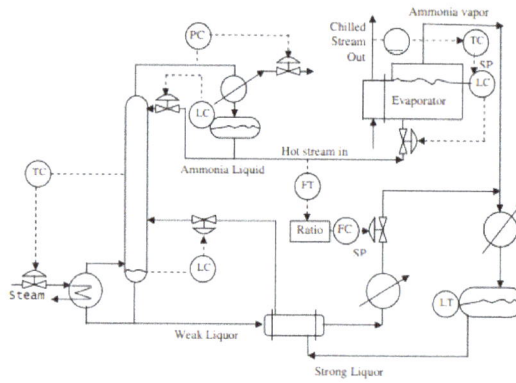

Absorption refrigeration controls

Control of Steam Utility System

Figure schematically shows a plant power and utility system. Boiler feed water is heated in a furnace to produce saturated steam. The saturated steam is passed through the furnace to produce superheated steam at 1000 psia pressure. The superheated LP steam drives a turbine to produce electricity. Steam at different pressures is extracted form the turbine for process steam utility requirements. Typically, steam at 300 psia (high pressure steam), 150 psia (medium pressure steam) and 50 psia (low pressure steam) is made available as a heat source at different temperature levels for process use. The pressure of the superheated steam from the furnace is maintained by adjusting the furnace duty. The steam drum level is controlled by adjusting the boiler feed water rate. The pressure of the 300 psia header is maintained in a split range arrangement as shown in the Figure. Steam from the higher pressure header is let in for a decrease in the header pressure while steam is dumped to the lower pressure header for an increase in the header pressure.

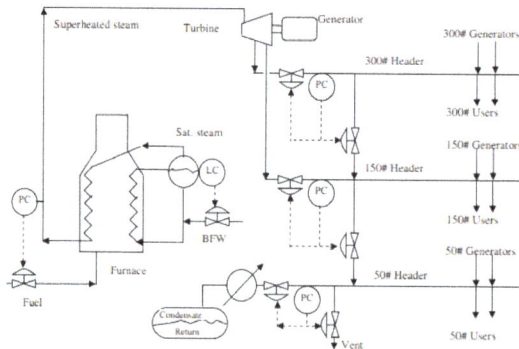

Power plant utility system controls

Plant-wide Modeling: An Overview

Mathematical model is a simulation of a scientific or industrial process. They can be very useful in the process industry when running actual processes are deemed too expensive. The topics discussed in the chapter are of great importance to broaden the existing knowledge on plant-wide modeling.

Process Modeling

A mathematical model of a process is a set of differential and algebraic equations whose solution yields the static and dynamic behavior of the process. In other words, a mathematical model essentially describes the physical and chemical phenomena of a process. Unarguably, performing experiments is the best way to gather information (data) about a process. However, time, effort and cost associated with experimental approach of gathering information may not be an easy option, especially for costly experiments. Mathematical model is a good alternative in such situations. However, it should be noted that a model is never a perfect alternative to the real life process. There is always a scope of error, however small it may be, between model and the process. Nevertheless, a reliable modelling exercise needs to be adopted so that the process/model mismatch remains as small as possible.

A mathematical model is useful for various purposes in the process industries. A trainee operator can learn the process details with the help of a process model. The operator needs to learn know-hows of critical situation and perform what-if analysis for the process. Such critical situation is never advisable "to be created" in the real-life plant operation. Hence, simulation of process model serves the purpose in this case. In fact the simulation of process model is carried out for design, safety analysis and controller synthesis of that process. Trial of the controller on a process model prior to the actual plant application is mandatory.

State Variables and Equations

The state variables are a set of fundamental dependent quantities whose values describe the natural state of a given process. The state equations are set of differential equations which describe the progression of the states with time. The state variables are primarily the fundamental quantities of a process viz., mass, energy and momentum, whereas the state equations are generated out of conservation principle involving these fundamental quantities.

Let S be the fundamental quantity. The state equation can be written as:

Rate of accumulation of S within the system

> = Flow rate of S coming to the system

> - Flow rate of S going of the system

> + Rate of S generated within the system

> - Rate of S consumed within the system

If there are N no. of streams going inside a process and M no. of streams coming out of a process, the mass balance equations for the process can be written as:

$$\frac{d(\rho V)}{dt} = \sum_{i=1}^{N} \rho_i F_i - \sum_{j=1}^{M} \rho_j F_j$$

Where ρ and V represent the density and volume of the material inside the system and $\rho_{\{i/j\}}$ and $F_{\{i/j\}}$ are the density and flow rate of the incoming/outgoing streams respectively. In case of a chemical reaction, let C_A and $C_{A\{i/j\}}$ be the concentrations of component A of material inside the process and the streams respectively and r be the rate of the reaction. The component balance equations can be written as

$$\frac{d(C_A V)}{dt} = \sum_{i=1}^{N} C_{A_i} F_i - \sum_{j=1}^{M} C_{A_j} F_j \pm rV$$

Let h and $h_{\{i/j\}}$ be the enthalpies of material inside the process and incoming/outgoing streams respectively, Q be the heat supplied to the system (or heat removed from the system), W is the work done on (or by) the system. Then the energy (E) balance equation for such system is written as

$$\frac{dE}{dt} = \sum_{i=1}^{N} \rho_i F_i h_i - \sum_{j=1}^{M} \rho_j F_j h_j \pm Q \pm W$$

As the chemical process plants are usually static, momentum balance equations are usually not required for such cases.

In addition to the above state equations, a few algebraic equations are also useful for modeling of a chemical process. Examples of such equations are:

Transport rate equation: $Q = UA(T_{steam} - T)$

Kinetic rate equation: $r = k_o e^{-\frac{E}{RT}} C_A$

Phase equilibrium equation: $y = kx$

Gas law:

$$PV = nRT$$

State Variable

A state variable is one of the set of variables that are used to describe the mathematical "state" of a dynamical system. Intuitively, the state of a system describes enough about the system to determine its future behaviour in the absence of any external forces affecting the system. Models that consist of coupled first-order differential equations are said to be in state-variable form.

Examples

- In mechanical systems, the position coordinates and velocities of mechanical parts are typical state variables; knowing these, it is possible to determine the future state of the objects in the system.

- In thermodynamics, a state variable is also called a state function. Examples include temperature, pressure, volume, internal energy, enthalpy, and entropy. In contrast heat and work are not state functions, but process functions.

- In electronic circuits, the voltages of the nodes and the currents through components in the circuit are usually the state variables.

- In ecosystem models, population sizes (or concentrations) of plants, animals and resources (nutrients, organic material) are typical state variables.

Control Systems Engineering

In control engineering and other areas of science and engineering, state variables are used to represent the states of a general system. The set of possible combinations of state variable values is called the state space of the system. The equations relating the current state of a system to its most recent input and past states are called the state equations, and the equations expressing the values of the output variables in terms of the state variables and inputs are called the output equations. As shown below, the state equations and output equations for a linear time invariant system can be expressed using coefficient matrices:

$$A \in RN^*N, \ B \in RN^*L, \ C \in RM^*N, \ D \in RM^*L,$$

where N, L and M are the dimensions of the vectors describing the state, input and output, respectively.

Discrete-time Systems

The state vector (vector of state variables) representing the current state of a discrete-time system (i.e. digital system) is $x[n]$, where n is the discrete point in time at which the system is being evaluated. The discrete-time state equations are

$$x[n+1] = Ax[n] + Bu[n],$$

which describes the next state of the system ($x[n+1]$) with respect to current state and inputs $u[n]$ of the system. The output equations are

$$y[n] = Cx[n] + Du[n],$$

which describes the output $y[n]$ with respect to current states and inputs $u[n]$ to the system.

Continuous Time Systems

The state vector representing the current state of a continuous-time system (i.e. analog system) is $x(t)$, and the continuous-time state equations giving the evolution of the state vector are

$$\frac{dx(t)}{dt} = Ax(t) + Bu(t),$$

which describes the continuous rate of change $\frac{dx(t)}{dt}$ of the state of the system with respect to current state x(t) and inputs u(t) of the system. The output equations are

$$y(t) = Cx(t) + Du(t),$$

which describes the output $y(t)$ with respect to current states $x(t)$ and inputs $u(t)$ to the system.

Example of Modeling a Stirred Tank Heater

Consider the stirred tank heater in the Figure as shown below. The question is what would change in case a change is occurred in the input condition (either in the manipulated variable or the disturbance). It is evident that inlet flow rate and its temperature are the input condition which can undergo a change and in such situation the mass and energy content (state variables) of the tank would show a progression. In normal situation, flow rate or temperature of an inlet flow does not have a potential to displace the tank physically from its normal position. Hence, there is no scope of progression of momentum of the tank. In other words, one need not carry out momentum balance operation on this process, rather mass and energy balance operation would suffice.

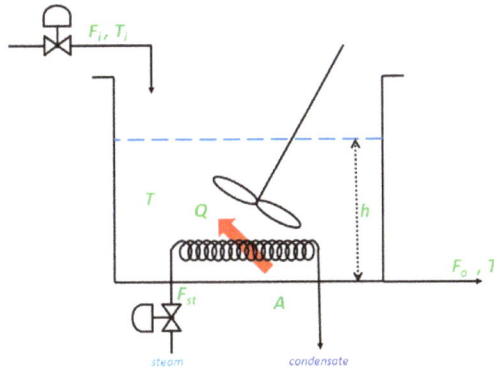

Schematic of a stirred tank heater

Let us now apply the material balance and energy balance operation on this process that would yield the following two equations:

Continuous Stirred-tank Reactor

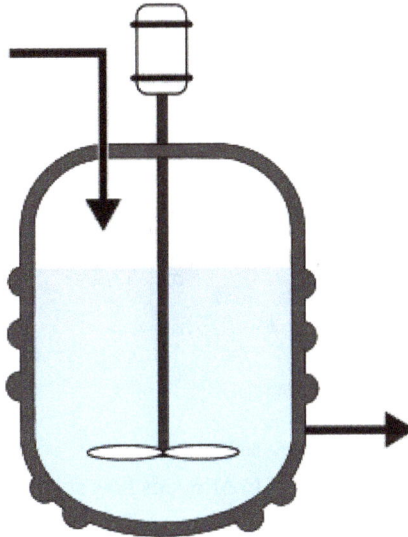

CSTR symbol

The continuous flow stirred-tank reactor (CSTR), also known as vat- or backmix reactor, is a common ideal reactor type in chemical engineering. A CSTR often refers to a model used to estimate the key unit operation variables when using a continuous[†] agitated-tank reactor to reach a specified output. The mathematical model works for all fluids: liquids, gases, and slurries.

The behavior of a CSTR is often approximated or modeled by that of a Continuous Ideally Stirred-Tank Reactor (CISTR). All calculations performed with CISTRs assume perfect mixing. In a perfectly mixed reactor, the output composition is identical to composition of the material inside the reactor, which is a function of residence time and rate of reaction. If the residence time is 5-10 times the mixing time, this approximation is valid for engineering purposes. The CISTR model is often used to simplify engineering calculations and can be used to describe research reactors. In practice it can only be approached, in particular in industrial size reactors.

Assume:

- perfect or ideal mixing, as stated above

Integral mass balance on number of moles N_i of species i in a reactor of volume V.

$$[\text{accumulation}] = [\text{in}] - [\text{out}] + [\text{generation}]$$

1. $$\frac{dN_i}{dt} = F_{io} - F_i + V v_i r_i$$

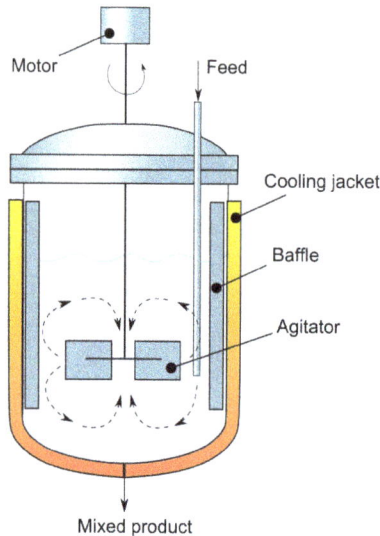

Cross-sectional diagram of Continuous flow stirred-tank reactor

where Fio is the molar flow rate inlet of species i, Fi the molar flow rate outlet, and stoichiometric coefficient. The reaction rate, r, is generally dependent on the reactant concentration and the rate constant (k). The rate constant can be determined by using a known empirical reaction rates that is adjusted for temperature using the Arrhenius temperature dependence. Generally, as the temperature increases so does the rate at which the reaction occurs. Residence time, τ, is the average amount of time a discrete quantity of reagent spends inside the tank.

Assume:

- constant density (valid for most liquids; valid for gases only if there is no net change in the number of moles or drastic temperature change)

- isothermal conditions, or constant temperature (k is constant)

- steady state ($G_A = r_A v$)

- single, irreversible reaction ($v_A = -1$)

- first-order reaction ($r = kC_A$)

$A \rightarrow$ products

$N_A = C_A V$ (where C_A is the concentration of species A, V is the volume of the reactor, N_A is the number of moles of species A)

2. $\qquad C_A = \dfrac{C_{Ao}}{1 + k\tau}$

The values of the variables, outlet concentration and residence time, in Equation 2 are major design criteria.

To model systems that do not obey the assumptions of constant temperature and a single reaction, additional dependent variables must be considered. If the system is considered to be in unsteady-state, a differential equation or a system of coupled differential equations must be solved.

CSTR's are known to be one of the systems which exhibit complex behavior such as steady-state multiplicity, limit cycles and chaos.

Application

Continuous flow stirred-tank reactors are usually applied in waste water treatment processes. CSTRs facilitate rapid dilution rates which make them resistant to both high pH and low pH volatile fatty acid wastes. CSTRs are less efficient compared to other types of reactors as they require larger reactor volumes to achieve the same reaction rate as other reactor models such as Plug Flow Reactors.

Mass Balance

A mass balance, also called a material balance, is an application of conservation of mass to the analysis of physical systems. By accounting for material entering and leaving a system, mass flows can be identified which might have been unknown, or difficult to measure without this technique. The exact conservation law used in the analysis of the system depends on the context of the problem, but all revolve around mass conservation, i.e. that matter cannot disappear or be created spontaneously.

Therefore, mass balances are used widely in engineering and environmental analyses. For example, mass balance theory is used to design chemical reactors, to analyse alternative processes to produce chemicals, as well as to model pollution dispersion and other processes of physical systems. Closely related and complementary analysis techniques include the population balance, energy balance and the somewhat more complex entropy balance. These techniques are required for thorough design and analysis of systems such as the refrigeration cycle.

In environmental monitoring the term budget calculations is used to describe mass balance equations where they are used to evaluate the monitoring data (comparing input and output, etc.) In biology the dynamic energy budget theory for metabolic organisation makes explicit use of mass and energy balances.

Introduction

The general form quoted for a mass balance is *The mass that enters a system must, by conservation of mass, either leave the system or accumulate within the system* .

Mathematically the mass balance for a system without a chemical reaction is as follows:

$$Input = Output + Accumulation$$

Strictly speaking the above equation holds also for systems with chemical reactions if the terms in the balance equation are taken to refer to total mass, i.e. the sum of all the chemical species of the system. In the absence of a chemical reaction the amount of any chemical species flowing in and out will be the same; this gives rise to an equation for each species present in the system. However, if this is not the case then the mass balance equation must be amended to allow for the generation or depletion (consumption) of each chemical species. Some use one term in this equation to account for chemical reactions, which will be negative for depletion and positive for generation. However, the conventional form of this equation is written to account for both a positive generation term (i.e. product of reaction) and a negative consumption term (the reactants used to produce the products). Although overall one term will account for the total balance on the system, if this balance equation is to be applied to an individual species and then the entire process, both terms are necessary. This modified equation can be used not only for reactive systems, but for population balances such as arise in particle mechanics problems. The equation is given below; note that it simplifies to the earlier equation in the case that the generation term is zero.

$$Input + Generation = Output + Accumulation + Consumption$$

- In the absence of a nuclear reaction the number of atoms flowing in and out must remain the same, even in the presence of a chemical reaction.

- For a balance to be formed, the boundaries of the system must be clearly defined.

- Mass balances can be taken over physical systems at multiple scales.

- Mass balances can be simplified with the assumption of steady state, in which the accumulation term is zero.

Illustrative Example

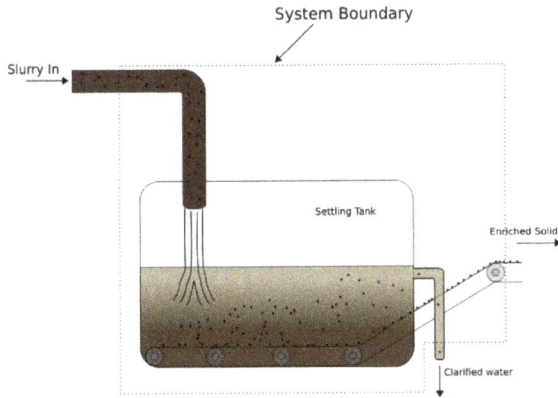

Diagram showing clarifier example

A simple example can illustrate the concept. Consider the situation in which a slurry is flowing into a settling tank to remove the solids in the tank. Solids are collected at the bottom by means of a conveyor belt partially submerged in the tank, and water exits via an overflow outlet.

In this example, there are two substances: solids and water. The water overflow outlet carries an increased concentration of water relative to solids, as compared to the slurry inlet, and the exit of the conveyor belt carries an increased concentration of solids relative to water.

Assumptions

- Steady state

- Non-reactive system

Analysis

Suppose that the slurry inlet composition (by mass) is 50% solid and 50% water, with a mass flow of 100 kg/min. The tank is assumed to be operating at steady state, and as such accumulation is zero, so input and output must be equal for both the solids and water. If we know that the removal efficiency for the slurry tank is 60%, then the water outlet will contain 20 kg/min of solids (40% times 100 kg/min times 50% solids). If we measure the flow rate of the combined solids and water, and the water outlet is shown to be 65 kg/min, then the amount of water exiting via the conveyor belt must be 10 kg/min. This allows us to completely determine how the mass has been distributed in the system with only limited information and using the mass balance relations across the system boundaries.

Mass Feedback (Recycle)

Cooling towers are a good example of a recycle system

Mass balances can be performed across systems which have cyclic flows. In these systems output streams are fed back into the input of a unit, often for further reprocessing.

Such systems are common in grinding circuits, where grain is crushed then sieved to only allow fine particles out of the circuit and the larger particles are returned to the roller mill (grinder). However, recycle flows are by no means restricted to solid mechanics operations; they are used in liquid and gas flows, as well. One such example is in cooling towers, where water is pumped through a tower many times, with only a small quantity of water drawn off at each pass (to prevent solids build up) until it has either evaporated or exited with the drawn off water.

The use of the recycle aids in increasing overall conversion of input products, which is useful for low per-pass conversion processes (such as the Haber process).

Differential Mass Balances

A mass balance can also be taken differentially. The concept is the same as for a large mass balance, but it is performed in the context of a limiting system (for example, one can consider the limiting case in time or, more commonly, volume). A differential mass balance is used to generate differential equations that can provide an effective tool for modelling and understanding the target system.

The differential mass balance is usually solved in two steps: first, a set of governing differential equations must be obtained, and then these equations must be solved, either analytically or, for less tractable problems, numerically.

The following systems are good examples of the applications of the differential mass balance:

1. Ideal (stirred) Batch reactor

2. Ideal tank reactor, also named Continuous Stirred Tank Reactor (CSTR)

3. Ideal Plug Flow Reactor (PFR)

Ideal Batch Reactor

The ideal completely mixed batch reactor is a closed system. Isothermal conditions are assumed, and mixing prevents concentration gradients as reactant concentrations decrease and product concentrations increase over time. Many chemistry textbooks implicitly assume that the studied system can be described as a batch reactor when they write about reaction kinetics and chemical equilibrium. The mass balance for a substance A becomes

$$IN + PROD = OUT + ACC$$

$$0 + r_A V = 0 + \frac{dn_A}{dt}$$

where r_A denotes the rate at which substance A is produced, V is the volume (which may be constant or not), n_A the number of moles (n) of substance A.

In a fed-batch reactor some reactants/ingredients are added continuously or in pulses (compare making porridge by either first blending all ingredients and then letting it boil, which can be described as a batch reactor, or by first mixing only water and salt and making that boil before the other ingredients are added, which can be described as a fed-batch reactor). Mass balances for fed-batch reactors become a bit more complicated.

Reactive Example

In the first example, we will show how to use a mass balance to derive a relationship between the percent excess air for the combustion of a hydrocarbon-base fuel oil and the percent oxygen in the combustion product gas. First, normal dry air contains 0.2095 mol of oxygen per mole of air, so there is one mole of O_2 in 4.773 mol of dry air. For stoichiometric combustion, the relationships between the mass of air and the mass of each combustible element in a fuel oil are:

$$\text{Carbon:} \quad \frac{mass\ of\ air}{mass\ of\ C} = \frac{4.773 \times 28.96}{12.01} = 11.51$$

$$\text{Hydrogen:} \quad \frac{mass\ of\ air}{mass\ of\ H} = \frac{\frac{1}{4}(4.773) \times 28.96}{1.008} = 34.28$$

$$\text{Sulfur:} \quad \frac{mass\ of\ air}{mass\ of\ S} = \frac{4.773 \times 28.96}{32.06} = 4.31$$

Considering the accuracy of typical analytical procedures, an equation for the mass of air per mass of fuel at stoichiometric combustion is:

$$\frac{mass\ of\ air}{mass\ of\ fuel} = AFR_{mass} = 11.5(wC) + 34.3(wH) + (wS - wO)$$

where wC, wH, wS, and wO refer to the mass fraction of each element in the fuel oil, sulfur burning to SO_2, and AFR_{mass} refers to the air-fuel ratio in mass units.

For 1 kg of fuel oil containing 86.1% C, 13.6% H, 0.2% O, and 0.1% S the stoichiometric mass of air is 14.56 kg, so AFR = 14.56. The combustion product mass is then 15.56 kg. At exact stoichiometry, O_2 should be absent. At 15 percent excess air, the AFR = 16.75, and the mass of the combustion product gas is 17.75 kg, which contains 0.505 kg of excess oxygen. The combustion gas thus contains 2.84 percent O_2 by mass. The relationships between percent excess air and %O_2 in the combustion gas are accurately expressed by quadratic equations, valid over the range 0–30 percent excess air:

$$\% \ excess\ air = 1.2804 \times (\%O_2\ in\ combustion\ gas)^2 + 4.49 \times (\%O_2\ in\ combustion\ gas)$$

$$\%O_2\ in\ combustion\ gas = -0.00138 \times (\%\ excess\ air)^2 + 0.210 \times (\%\ excess\ air)$$

In the second example we will use the law of mass action to derive the expression for a chemical equilibrium constant.

Assume we have a closed reactor in which the following liquid phase reversible reaction occurs:

$$aA + bB \leftrightarrow cC + dD$$

The mass balance for substance A becomes

$$IN + PROD = OUT + ACC$$

$$0 + r_A V = 0 + \frac{dn_A}{dt}$$

As we have a liquid phase reaction we can (usually) assume a constant volume and since $n_A = V * C_A$ we get

$$r_A V = V \frac{dC_A}{dt}$$

or

$$r_A = \frac{dC_A}{dt}$$

In many textbooks this is given as the definition of reaction rate without specifying the implicit assumption that we are talking about reaction rate in a closed system with only one reaction. This is an unfortunate mistake that has confused many students over the years.

According to the law of mass action the forward reaction rate can be written as

$$r_1 = k_1 [A]^a [B]^b$$

and the backward reaction rate as

$$r_{-1} = k_{-1} [C]^c [D]^d$$

The rate at which substance A is produced is thus

$$r_A = a(r_{-1} - r_1)$$

and since, at equilibrium, the concentration of A is constant we get

$$r_A = a(r_{-1} - r_1) = \frac{dC_A}{dt} = 0$$

or, rearranged

$$\frac{k_1}{k_{-1}} = \frac{[C]^c [D]^d}{[A]^a [B]^b} = K_{eq}$$

Ideal Tank Reactor/Continuously Stirred Tank Reactor

The continuously mixed tank reactor is an open system with an influent stream of reactants and an effluent stream of products. A lake can be regarded as a tank reactor, and lakes with long turnover times (e.g. with low flux-to-volume ratios) can for many purposes be regarded as continuously stirred (e.g. homogeneous in all respects). The mass balance then becomes

$$IN + PROD = OUT + ACC$$

$$Q_0 \cdot C_{A,0} + r_A \cdot V = Q \cdot C_A + \frac{dn_A}{dt}$$

where Q_0 and Q denote the volumetric flow in and out of the system respectively and $C_{A,0}$ and C_A the concentration of A in the inflow and outflow respective. In an open system we can never reach a chemical equilibrium. We can, however, reach a steady state where all state variables (temperature, concentrations etc.) remain constant ($ACC = 0$).

Example

Consider a bathtub in which there is some bathing salt dissolved. We now fill in more water, keeping the bottom plug in. What happens?

Since there is no reaction, $PROD = 0$ and since there is no outflow $Q = 0$. The mass balance becomes

$$IN + PROD = OUT + ACC$$

$$Q_0 \cdot C_{A,0} + 0 = 0 \cdot C_A + \frac{dn_A}{dt}$$

or

$$Q_0 \cdot C_{A,0} = \frac{dC_A V}{dt} = V \frac{dC_A}{dt} + C_A \frac{dV}{dt}$$

Using a mass balance for total volume, however, it is evident that $\frac{dV}{dt} = Q_0$ and that $V = V_{t=0} + Q_0 t$. Thus we get

$$\frac{dC_A}{dt} = \frac{Q_0}{(V_{t=0} + Q_0 t)}\left(C_{A,0} - C_A\right)$$

Note that there is no reaction and hence no reaction rate or rate law involved, and yet $\frac{dC_A}{dt} \neq 0$. We can thus draw the conclusion that reaction rate can not be defined in a general manner using $\frac{dC}{dt}$. One must first write down a mass balance before a link between $\frac{dC}{dt}$ and the reaction rate can be found. Many textbooks, however, define reaction rate as

$$v = \frac{dC_A}{dt}$$

without mentioning that this definition implicitly assumes that the system is closed, has a constant volume and that there is only one reaction.

Ideal Plug Flow Reactor (PFR)

The idealized plug flow reactor is an open system resembling a tube with no mixing in the direction of flow but perfect mixing perpendicular to the direction of flow. Often used for systems like rivers and water pipes if the flow is turbulent. When a mass balance is made for a tube, one first considers an infinitesimal part of the tube and make a mass balance over that using the ideal tank reactor model. That mass balance is then integrated over the entire reactor volume to obtain:

$$\frac{d(Q \cdot C_A)}{dV} = r_A$$

In numeric solutions, e.g. when using computers, the ideal tube is often translated to a series of tank reactors, as it can be shown that a PFR is equivalent to an infinite number of stirred tanks in series, but the latter is often easier to analyze, especially at steady state.

More Complex Problems

In reality, reactors are often non-ideal, in which combinations of the reactor models above are used to describe the system. Not only chemical reaction rates, but also mass transfer rates may be important in the mathematical description of a system, especially in heterogeneous systems.

As the chemical reaction rate depends on temperature it is often necessary to make both an energy balance (often a heat balance rather than a full-fledged energy balance) as well as mass balances to fully describe the system. A different reactor model might be needed for the energy balance: A system that is closed with respect to mass might be open with respect to energy e.g. since heat may enter the system through conduction.

Commercial Use

In industrial process plants, using the fact that the mass entering and leaving any portion of a process plant must balance, data validation and reconciliation algorithms may be employed to correct measured flows, provided that enough redundancy of flow measurements exist to permit statistical reconciliation and exclusion of detectably erroneous measurements. Since all real world measured values contain inherent error, the reconciled measurements provide a better basis than the measured values do for financial reporting, optimization, and regulatory reporting. Software packages exist to make this commercially feasible on a daily basis.

Material Balance

Rate of accumulation of water = rate of water inlet - rate of water outlet

$$\frac{dm}{dt} = m_i - m_o$$

$$\frac{d}{dt}\{\rho V\} = \frac{d}{dt}\{\rho Ah\} = F_i \rho_i - F_o \rho_o$$

Where A is the cross sectional area of the tank. If we assume the density of water to be constant then the material balance equation would take the final form as

$$A\frac{dh}{dt} = F_i - F_o$$

For a free flow system,

$$F_o = c\sqrt{h}$$

Where c is a constant. Hence,

$$A\frac{dh}{dt} = F_i - c\sqrt{h}$$

Energy Balance

Rate of accumulation of heat = rate of heat in - rate of heat out + rate of heat supplied

$$\frac{d}{dt}\{mc_p \Delta T\} = m_i c_{p_i} \Delta T_i - m_o c_{p_o} \Delta T_o + Q$$

$$\frac{d}{dt}\{(\rho Ah)c_p (T - T_{ref})\} = (F_i \rho_i)c_{p_i}(T_i - T_{ref}) - (F_o \rho_o)c_{p_o}(T - T_{ref}) + Q$$

If we assume the density and specific heat of water to be constant and the reference temperature to be zero, then the energy balance equation would take the form as

$$A\frac{d}{dt}\{hT\} = F_i T_i - F_o T + \frac{Q}{\rho c_p}$$

$$Ah\frac{dT}{dt} + AT\frac{dh}{dt} = F_i T_i - F_o T + \frac{Q}{\rho c_p}$$

$$Ah\frac{dT}{dt}+T\left(F_i-F_o\right)=F_iT_i-F_oT+\frac{Q}{\rho c_p}$$

$$Ah\frac{dT}{dt}=F_i\left(T_i-T\right)+\frac{Q}{\rho c_p}$$

Equations represent the mathematical model of the stirred tank heater.

State Variables	$: h, T$	
Output Variables	$: h, T$	(*both measured*)
Input Variables	$: Q\left(orF_{st}\right), F_i$	(*manipulated*)
	$: T_i$	(*disturbance*)
Constant Parameters	$: c, A, \rho, c_\rho$	

Degrees of Freedom

The degree of freedom is defined by the total number. of independent variables that must be specified in order to define the system completely. In other words, it is the minimum number. of independent variables required to be specified so that the model equation(s) can be solved.

Let us analyze the case of the stirred tank heater:

Total no. of variables	$: 5$	(h, T, F_i, T_i, Q)
Total no. of equations	$: 2$	

With only two equations in hand, one can solve for only two unknown variables in order to obtain a unique solution. Hence, out of 5 variables, 3 needs to be specified before an unique solution of the equation can be attempted. It is evident from the list of variables that one needs to specify F_i, T_i and Q in order to solve for the state variables h and T . Essentially the degree of freedom is calculated by subtracting the number of equations from number. of variables. Thus the degree of freedom of the stirred tank heater process is 3.

Simulation of the Mathematical Model of Stirred Tank Heater

The model consists of two ordinary differential equations. Simulation of the model (i.e. solving these model equations) with an "input condition" yields the data along with some percentage of error, which are otherwise expected, from the real-time process operation at same "input condition". Although equations are simple first order ODEs, the nonlinear terms of those model equations may be hindrance against getting analytical solution of the model. Numerical solution, on the other hand, is a popular approach for simulation of such model equations. Various simulation software are available for easy use, e.g. MATLAB and SIMULINK to name a few. Following subsection describes the model of stirred tank heater developed in SIMULINK domain.

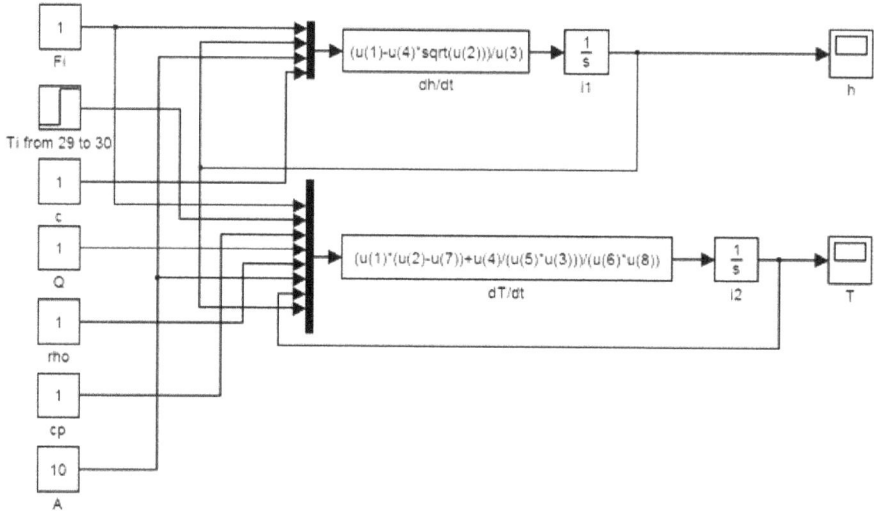

Model of stirred tank heater developed in SIMULINK domain

In order to solve the model equation one needs to specify the model parameters and satisfy the degree of freedom. Let us consider the following numerical values for the said parameters and variables: $A = 10$; $c = 1$; $c_p = 1$; $\rho = 1$; $F_i = 1$; $T_i = 29$; $Q = 1$. At steady state, left hand side of equations would be zero.

$$0 = F_i - c\sqrt{h} \qquad (II.8)$$

$$0 = F_i(T_i - T) + \frac{Q}{\rho c_p} \qquad (II.9)$$

Solution of equations yields the steady state values of the state variables: $h_s = 1$ and 30. The values of h and T would stay at 1 and 30 respectively as long as there is no change in the input condition.

Let us analyze what would happen if the temperature of inlet flow increases from 29 to 30. Solution of equations II.8 and II.9 with $T_i = 30$ yields a new steady state condition of state variables as $h_s = 1$ and $T_s = 31$. That means a disturbance in T_i would not change the height of the liquid inside the tank however it would increase the temperature of the liquid to an increased new steady state. The transition of T_s from 30 to 31 can be observed by simulating the model in SIMULINK domain.

Transient response of temperature of liquid on giving a step disturbance to inlet temp

It is observed from the figure that the process takes 70 s to reach the new steady state of the liquid temperature inside the stirred tank heater.

Linearization of Nonlinear Model

From Equation, we observe that the steady state value of $h = 1$ when the inlet flow rate is maintained at $F_i = 1$.

Case I: Inlet flow rate undergoes a step change and new value is $F_i = 2$. As a result the h would attain a new steady state value of 4. The gain of the system is defined as the ratio of change in output to change in input. Hence, $\text{gain}\left(K\right) = \dfrac{\Delta h_s}{\Delta F_{is}} = \dfrac{4-1}{2-1} = 3$.

Case II: Inlet flow rate undergoes a step change and new value is $F_i = 3$. As a result the h would attain a new steady state value of 9. Hence, $\text{gain}\left(K\right) = \dfrac{\Delta h_s}{\Delta F_{is}} = \dfrac{9-1}{3-1} = 4$.

It is observed that the gain of the system is not constant. The gain changes with various input conditions of the system. This is the identity of a nonlinear process. In general, if a system follows the *principle of superposition* then it is termed as linear process. Usually, a nonlinear process is identified with existence of a nonlinear term in its model such as logarithmic, power, exponential, product etc. Existence of square root term with h yields nonlinearity in the stirred tank heater process.

Most of significant developments in the control theories have occurred for linear processes as easy analytical solutions are available for them. Unfortunately, most of the chemical processes are nonlinear in nature. Figure pictorially describes the effect of linearization of a nonlinear process.

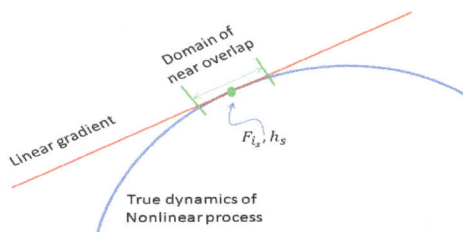

The domain of linearization

Suppose the blue line indicates the true dynamics of a nonlinear process. A steady state point $\{F_{i_s}, h_s\}$ is identified on this line that represents the nominal point of operation of the process. If the process is linearized in the neighborhood of the nominal operating point then the resulting gradient can be indicated by the red line. It is evident from the figure that there lies a considerable portion of domain in the neighborhood of the nominal point of operation over which the dynamics of both nonlinear and linearized process match. Hence, it can be argued that the basic aim of a control system is to maintain the operation of the process at a predefined nominal point. Even if a disturbance takes the process away from this nominal point, an ideal control system should ensure that the departure of the process from its nominal point of operation remains within the limit of such neighborhood and the process is eventually brought back at its nominal operating point within reasonable time. A linearized form of the model is likely to cover sufficient domain around the nominal operating point so that its dynamics reasonably matches with that of the actual nonlinear model. In such situation, it is customary to design a linear controller on the basis of linearized model and apply it on the nonlinear process.

The linearization of model is carried out by using Taylor series expansion. The expansion of a function $f(x)$ around x_0 is given as

$$f(x) = f(x_0) + \frac{df}{dx}\bigg|_{x_0} \frac{(x-x_0)}{1!} + \frac{d^2 f}{dx^2}\bigg|_{x_0} \frac{(x-x_0)^2}{2!} + \frac{d^3 f}{dx^3}\bigg|_{x_0} \frac{(x-x_0)^3}{3!} + \ldots\ldots$$

When the value of x is very close to x_0, then the power terms of $(x-x_0)$ are very small and hence can be neglected. The resulting function will have the form:

$$f(x) \approx f(x_0) + \frac{df}{dx}\bigg|_{x_0} \frac{(x-x_0)}{1!} \qquad\qquad II \quad a$$

The equation above is a linear function. The linearization of a function $f(x, y, \ldots)$ around $\{x_o, y_o, \ldots\}$ is given as

$$f(x, y, \ldots) = f(x_o, y_o, \ldots) + \frac{\partial f}{\partial x}\bigg|_{x_o y_o \ldots} \frac{(x-x_0)}{1!} + \frac{\partial f}{\partial y}\bigg|_{x_o y_o \ldots} \frac{(y-y_0)}{1!} + \ldots$$

Linearization of Stirred Tank Heater

Consider the equation of the stirred tank heater. Only nonlinear term in the equation is $\sqrt{h}$. Linearization of this nonlinear term would yield the following:

$$\sqrt{h} \approx \sqrt{h_s} + \frac{d(\sqrt{h})}{dh}\bigg|_{h_s} \frac{(h-h_s)}{1!} = \sqrt{h_s} + \frac{1}{2\sqrt{h_s}}(h-h_s)$$

Hence the linearized form of model equation would be

$$A\frac{dh}{dt} = F_i - c\sqrt{h} = F_i - c\left\{\sqrt{h_s} + \frac{1}{2\sqrt{h_s}}(h - h_s)\right\} = F_i - \left(\frac{c}{2\sqrt{h_s}}\right)h - \left(\frac{c\sqrt{h_s}}{2}\right)$$

Or

$$A\frac{dh}{dt} = F_i - \left(\frac{c}{2\sqrt{h_s}}\right)h - \left(\frac{c\sqrt{h_s}}{2}\right)$$

Deviation Variables

As the aim of the control system is to maintain the process at a nominal operating point, the deviation or dislocation of the state from its nominal point is a crucial variable to observe. The deviation variable of a state describes the departure of the state from its nominal point of operation.

Let us consider the linearized model of the stirred tank heater as given by Equation. At steady state,

$$A\frac{dh_s}{dt} = F_{i_s} - \left(\frac{c}{2\sqrt{h_s}}\right)h_s - \left(\frac{c\sqrt{h_s}}{2}\right)$$

Subtracting Equation from we obtain,

$$A\frac{d(h - h_s)}{dt} = \left(F_i - F_{i_s}\right) - \left(\frac{c}{2\sqrt{h_s}}\right)(h - h_s)$$

Or,

$$A\frac{d\bar{h}}{dt} = \bar{F}_i - \left(\frac{c}{2\sqrt{h_s}}\right)\bar{h}$$

Equation represents the deviation form of the model. The variables $\bar{h}$ and $\bar{F}_i$ represent the deviation variables of the height of liquid in the stirred tank heater and that of the inlet flow rate.

References

- Dales, Richard C. (1973), The Scientific Achievement of the Middle Ages (The Middle Ages Series), University of Pennsylvania Press, ISBN 9780812210576

- Einstein, Albert; Infeld, Leopold (1938), The Evolution of Physics: from early concepts to relativity and quanta, New York: Simon and Schuster, ISBN 0-671-20156-5

- Goldhaber, Alfred Scharff; Nieto, Michael Martin (January–March 2010), "Photon and graviton mass limits", Rev. Mod. Phys., American Physical Society, 82: 939, doi:10.1103/RevModPhys.82.939 . pp. 939–79

- Glen, William (ed.) (1994), The Mass-Extinction Debates: How Science Works in a Crisis, Stanford, CA: Stanford University Press, ISBN 0-8047-2285-4

- Newton, Isaac (1999) [1687, 1713, 1726], Philosophiae Naturalis Principia Mathematica, University of California Press, ISBN 0-520-08817-4

 4

Instrumentation in Chemical Plants

Instrumentation is the term used for measuring and indicating physical quantity. Simple examples of instrumentation systems are sensors and the mechanical thermostat. This section discusses the methods of instrumentation in a critical manner providing key analysis to the subject matter.

Instrumentation

The instrumentation part refers mainly to the hardware of a control systems that includes various measuring instruments, transmitters, valves etc. In this chapter, some basic knowledge on instrumentation symbols as well as process drawings will be discussed.

Instrumentation is a collective term for measuring instruments used for indicating, measuring and recording physical quantities.

The term instrumentation may refer to something as simple as direct reading thermometers or, when using many sensors, may become part of a complex Industrial control system in such as manufacturing industry, vehicles and transportation. Instrumentation can be found in the household as well; a smoke detector or a heating thermostat are examples.

History and Development

A local instrumentation panel on a steam turbine.

The history of instrumentation can be divide into several phases.

Pre-industrial

Elements of industrial instrumentation have long histories. Scales for comparing weights and simple pointers to indicate position are ancient technologies. Some of the earliest measurements were of time. One of the oldest water clocks was found in the tomb of the ancient Egyptian pharaoh Amenhotep I, buried around 1500 BCE. Improvements were incorporated in the clocks. By 270 BCE they had the rudiments of an automatic control system device.

In 1663 Christopher Wren presented the Royal Society with a design for a "weather clock". A drawing shows meteorological sensors moving pens over paper driven by clockwork. Such devices did not become standard in meteorology for two centuries. The concept has remained virtually unchanged as evidenced by pneumatic chart recorders, where a pressurized bellows displaces a pen. Integrating sensors, displays, recorders and controls was uncommon until the industrial revolution, limited by both need and practicality.

Early Industrial

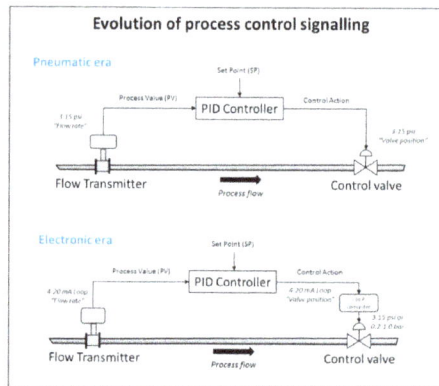

The evolution of analogue control loop signalling
from the pneumatic era to the electronic era.

Early systems used direct process connections the control panels for indications. As the systems became larger, the potential for blocked piping and other hazards related to toxic materials led to the introduction of pneumatic transmitters and then to automatic 3-term (PID) controllers in the late 1930's. The first of these controllers,, produced by Honeywell, were used in volume by the Manhattan Project in the processing of uranium compounds. The ranges of pneumatic transmitters were defined by the need to control valves and actuators in the field. Typically a signal ranged from 3 to 15 psi (20 to 100kPa or 0.2 to 1.0 kg/cm^2) as a standard, was standardized with 6 to 30 psi occasionally being used for larger valves. Transistor electronics enabled wiring to replace pipes, initially with a range of 20 to 100mA at up to 90V for loop powered devices, reducing to

4 to 20mA at 12 to 24V in more modern systems. A transmitter is a device that produces an output signal, often in the form of a 4–20 mA electrical current signal, although many other options using voltage, frequency, pressure, or ethernet are possible. The transistor was commercialized by the mid-1950s.

Instruments attached to a control system provided signals used to operate solenoids, valves, regulators, circuit breakers, relays and other devices. Such devices could control a desired output variable, and provide either remote or automated control capabilities.

Each instrument company introduced their own standard instrumentation signal, causing confusion until the 4-20 mA range was used as the standard electronic instrument signal for transmitters and valves. This signal was eventually standardized as ANSI/ISA S50, "Compatibility of Analog Signals for Electronic Industrial Process Instruments", in the 1970s. The transformation of instrumentation from mechanical pneumatic transmitters, controllers, and valves to electronic instruments reduced maintenance costs as electronic instruments were more dependable than mechanical instruments. This also increased efficiency and production due to their increase in accuracy. Pneumatics enjoyed some advantages, being favored in corrosive and explosive atmospheres.

Automatic Process Control

In the early years of process control, process indicators and control elements such as valves were monitored by an operator that walked around the unit adjusting the valves to obtain the desired temperatures, pressures, and flows. As technology evolved pneumatic controllers were invented and mounted in the field that monitored the process and controlled the valves. This reduced the amount of time process operators were needed to monitor the process. Later years the actual controllers were moved to a central room and signals were sent into the control room to monitor the process and outputs signals were sent to the final control element such as a valve to adjust the process as needed. These controllers and indicators were mounted on a wall called a control board. The operators stood in front of this board walking back and forth monitoring the process indicators. This again reduced the number and amount of time process operators were needed to walk around the units. The most standard pneumatic signal level used during these years was 3-15 psig.

Large Integrated Computer-based Systems

Process control of large industrial plants has evolved through many stages. Initially, control would be from panels local to the process plant. However this required a large manpower resource to attend to these dispersed panels, and there was no overall view of the process. The next logical development was the transmission of all plant measurements to a permanently-manned central control room. Effectively this was the centralisation of all the localised panels, with the advantages of lower manning levels and easier overview of the process. Often the controllers were behind the control room panels, and all automatic and manual control outputs were transmitted back to plant.

Pneumatic "Three term" pneumatic PID controller, widely used before electronics became reliable and cheaper and safe to use in hazardous areas (Siemens Telepneu Example)

A pre-DCS/SCADA era central control room. Whilst the controls are centralised in one place, they are still discrete and not integrated into one system.

However, whilst providing a central control focus, this arrangement was inflexible as each control loop had its own controller hardware, and continual operator movement within the control room was required to view different parts of the process. With coming of electronic processors and graphic displays it became possible to replace these discrete controllers with computer-based algorithms, hosted on a network of input/output racks with their own control processors. These could be distributed around plant, and communicate with the graphic display in the control room or rooms. The distributed control concept was born.

A DCS control room where plant information and controls are displayed on computer graphics screens. The operators are seated and can view and control any part of the process from their screens, whilst retaining a plant overview.

The introduction of DCSs and SCADA allowed easy interconnection and re-configuration of plant controls such as cascaded loops and interlocks, and easy interfacing with other production computer systems. It enabled sophisticated alarm handling, introduced automatic event logging, removed the need for physical records such as chart recorders, allowed the control racks to be networked and thereby located locally to plant to reduce cabling runs, and provided high level overviews of plant status and production levels.

Applications

In some cases the sensor is a very minor element of the mechanism. Digital cameras and wristwatches might technically meet the loose definition of instrumentation because they record and/or display sensed information. Under most circumstances neither would be called instrumentation, but when used to measure the elapsed time of a race and to document the winner at the finish line, both would be called instrumentation.

Household

A very simple example of an instrumentation system is a mechanical thermostat, used to control a household furnace and thus to control room temperature. A typical unit senses temperature with a bi-metallic strip. It displays temperature by a needle on the free end of the strip. It activates the furnace by a mercury switch. As the switch is rotated by the strip, the mercury makes physical (and thus electrical) contact between electrodes.

Another example of an instrumentation system is a home security system. Such a system consists of sensors (motion detection, switches to detect door openings), simple algorithms to detect intrusion, local control (arm/disarm) and remote monitoring of the system so that the police can be summoned. Communication is an inherent part of the design.

Kitchen appliances use sensors for control.

- A refrigerator maintains a constant temperature by measuring the internal temperature.

- A microwave oven sometimes cooks via a heat-sense-heat-sense cycle until sensing done.

- An automatic ice machine makes ice until a limit switch is thrown.

- Pop-up bread toasters can operate by time or by heat measurements.

- Some ovens use a temperature probe to cook until a target internal food temperature is reached.

- A common toilet refills the water tank until a float closes the valve. The float is acting as a water level sensor.

Automotive

Modern automobiles have complex instrumentation. In addition to displays of engine rotational speed and vehicle linear speed, there are also displays of battery voltage and current, fluid levels, fluid temperatures, distance traveled and feedbacks of various controls (turn signals, parking brake, headlights, transmission position). Cautions may be displayed for special problems (fuel low, check engine, tire pressure low, door ajar, seat belt unfastened). Problems are recorded so they can be reported to diagnostic equipment. Navigation systems can provide voice commands to reach a destination. Automotive instrumentation must be cheap and reliable over long periods in harsh environments. There may be independent airbag systems which contain sensors, logic and actuators. Anti-skid braking systems use sensors to control the brakes, while cruise control affects throttle position. A wide variety of services can be provided via communication links as the OnStar system. Autonomous cars (with exotic instrumentation) have been demonstrated.

Aircraft

Early aircraft had a few sensors. "Steam gauges" converted air pressures into needle deflections that could be interpreted as altitude and airspeed. A magnetic compass provided a sense of direction. The displays to the pilot were as critical as the measurements.

A modern aircraft has a far more sophisticated suite of sensors and displays, which are embedded into avionics systems. The aircraft may contain inertial navigation systems, global positioning systems, weather radar, autopilots, and aircraft stabilization systems. Redundant sensors are used for reliability. A subset of the information may be transferred to a crash recorder to aid mishap investigations. Modern pilot displays now include computer displays including head-up displays.

Air traffic control radar is distributed instrumentation system. The ground portion transmits an electromagnetic pulse and receives an echo (at least). Aircraft carry transponders that transmit codes on reception of the pulse. The system displays aircraft map location, an identifier and optionally altitude. The map location is based on sensed antenna direction and sensed time delay. The other information is embedded in the transponder transmission.

Laboratory Instrumentation

Among the possible uses of the term is a collection of laboratory test equipment controlled by a computer through an IEEE-488 bus (also known as GPIB for General Purpose Instrument Bus or HPIB for Hewlitt Packard Instrument Bus). Laboratory equipment is available to measure many electrical and chemical quantities. Such a collection of equipment might be used to automate the testing of drinking water for pollutants.

Measurement Parameters

Instrumentation is used to measure many parameters (physical values). These parameters include:

• Pressure, either differential or static • Flow • Temperature • Levels of liquids, etc. • Density	• Viscosity • Ionising radiation • Frequency • Current	• Voltage • Inductance • Capacitance • Resistivity	• Chemical composition • Chemical properties • Vibration • Weight

Control valve.

Instrumentation Engineering

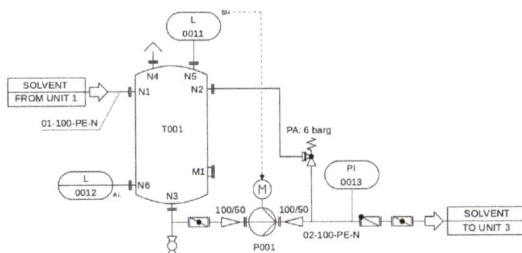

The instrumentation part of a Piping and instrumentation diagram
will be developed by an instrumentation engineer.

Instrumentation engineering is the engineering specialization focused on the principle and operation of measuring instruments that are used in design and configuration of automated systems in electrical, pneumatic domains etc. They typically work for industries with automated processes, such as chemical or manufacturing plants, with the goal of improving system productivity, reliability, safety, optimization, and stability. To control the parameters in a process or in a particular system, devices such as microprocessors, microcontrollers or PLCs are used, but their ultimate aim is to control the parameters of a system.

Instrumentation engineering is loosely defined because the required tasks are very domain dependent. An expert in the biomedical instrumentation of laboratory rats has very different concerns than the expert in rocket instrumentation. Common concerns of both are the selection of appropriate sensors based on size, weight, cost, reliability, accuracy, longevity, environmental robustness and frequency response. Some sensors are literally fired in artillery shells. Others sense thermonuclear explosions until destroyed. Invariably sensor data must be recorded, transmitted or displayed. Recording rates and capacities vary enormously. Transmission can be trivial or can be clandestine, encrypted and low-power in the presence of jamming. Displays can be trivially simple or can require consultation with human factors experts. Control system design varies from trivial to a separate specialty.

Instrumentation engineers are responsible for integrating the sensors with the recorders, transmitters, displays or control systems, and producing the Piping and instrumentation diagram for the process. They may design or specify installation, wiring and signal conditioning. They may be responsible for calibration, testing and maintenance of the system.

In a research environment it is common for subject matter experts to have substantial instrumentation system expertise. An astronomer knows the structure of the universe and a great deal about telescopes - optics, pointing and cameras (or other sensing elements). That often includes the hard-won knowledge of the operational procedures that provide the best results. For example, an astronomer is often knowledgeable of techniques to minimize temperature gradients that cause air turbulence within the telescope.

Instrumentation technologists, technicians and mechanics specialize in troubleshooting, repairing and maintaining instruments and instrumentation systems.

Typical Industrial Transmitter Signal Types

Current Loop (4-20mA) - Electrical

HART - Data signalling often overlaid on a current loop.

Foundation Fieldbus - Data signalling

Profibus - Data signalling

Impact of Modern Development

Ralph Müller (1940) stated "That the history of physical science is largely the history of instruments and their intelligent use is well known. The broad generalizations and theories which have arisen from time to time have stood or fallen on the basis of accurate measurement, and in several instances new instruments have had to be devised for the purpose. There is little evidence to show that the mind of modern man is superior to that of the ancients. His tools are incomparably better."

Davis Baird has argued that the major change associated with Floris Cohen's identification of a "fourth big scientific revolution" after World War II is the development of scientific instrumentation, not only in chemistry but across the sciences. In chemistry, the introduction of new instrumentation in the 1940s was "nothing less than a scientific and technological revolution" in which classical wet-and-dry methods of structural organic chemistry were discarded, and new areas of research opened up.

As early as 1954, W A Wildhack discussed both the productive and destructive potential inherent in process control. The ability to make precise, verifiable and reproducible measurements of the natural world, at levels that were not previously observable, using scientific instrumentation, has "provided a different texture of the world". This instrumentation revolution fundamentally changes human abilities to monitor and respond, as is illustrated in the examples of DDT monitoring and the use of UV spectrophotometry and gas chromatography to monitor water pollutants.

Piping and Instrumentation Diagram (P&ID)

Instrumentation Symbols

Symbols of standardized instruments are given in the following figure.

	Accessible to operator Primary	Field location	Accessible to operator Secondary	Inaccessible to operator
Discrete Instruments				
Shared display or control				
Computer Function				
PLC				

Standardized Instruments Symbols

Each instrument is identified with a letter of English alphabet. Table refers to the identification letters

Table: Instruments Identification Letters

| First letter + Modifier | | Succeeding letters | | |
Initiating or measured variable	Modifier	Readout or passive function	Output function	Modifier
A　Analysis		Alarm		
B　Burner, combustion		User's choice	User's choice	User's choice
C　User's choice			Control	
D　User's choice	Differential			
E　Voltage		Sensor		
F　Flow rate	Ratio			
G　User's choice		Glass, viewing device		
H　Hand				High
I　Current		Indicate		
J　Power	Scan			
K　Time	Time rate of change		Control station	
L　Level		Light		Low
M　User's choice	Momentary			Middle
N　User's choice		User's choice	User's choice	User's choice
O　User's choice		Orifice		
P　Pressure		Test point		
Q　Quantity	Integrate, totalize			
R　Radiation		Record		
S　Speed, frequency	Safety		Switch	
T　Temperature			Transmit	
U　Multivariable		Multifunction	Multifunction	Multifunction
V　Vibration, mechanical analysis			Valve, damper, louver	
W　Weight, force		Well		
X　Unclassified	x-axis	Unclassified	Unclassified	Unclassified
Y　Event, state, or presence	y-axis		Ready, compute, convert	
Z　Position, dimension	z-axis		Driver, actuator	

figure represents a few examples of the letters and numbering code that can be used to refer the instruments. Consider the example TY178 which has two letters T & Y . Table refers to the first letter T as temperature and second letter Y as converter. Hence it is a temperature transducer which converts a 4-20 mA current signal into a 3-15 psi pressure signal. The number 178 refers to the location of the transducer such as "zone 1, equipment number. 7, transducer number. 8". From Figure, we can further infer that transducer is a discrete instrument located in the field itself.

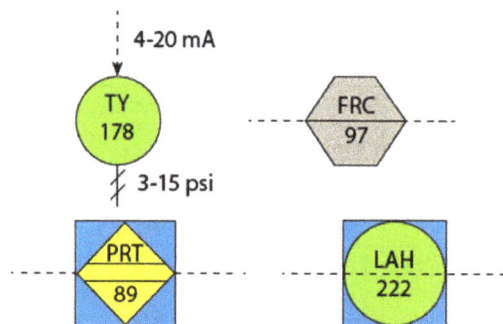

Examples of the letters and numbering code

Symbols can be broadly classified into four types depending on their functions:

- Actuator

- Primary elements

- Regulators and safety valves

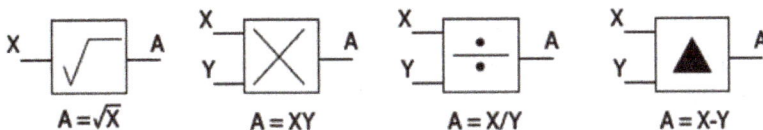

- Math function

Following are the examples of those categories

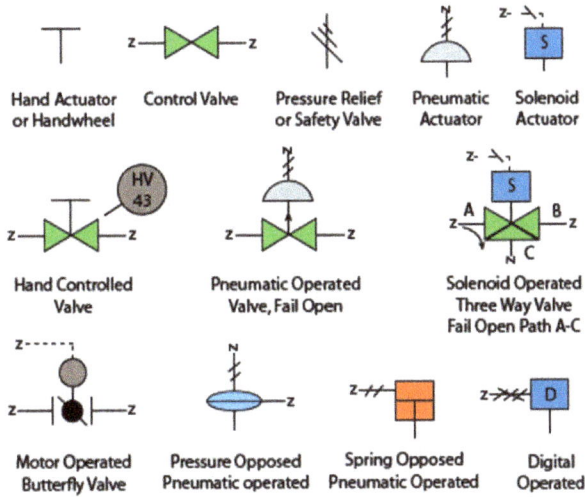

Hand Actuator   Control Valve   Pressure Relief   Pneumatic   Solenoid
or Handwheel   or Safety Valve   Actuator   Actuator

Hand Controlled   Pneumatic Operated   Solenoid Operated
Valve   Valve, Fail Open   Three Way Valve
                                               Fail Open Path A-C

Motor Operated   Pressure Opposed   Spring Opposed   Digital
Butterfly Valve   Pneumatic operated   Pneumatic Operated   Operated

Actuators

Orifice Plate with Vena   Turbine or Propeller   Magnetic Flowmeter   Bimetalic, Glass, or   Thermal - Radiation   Speed Transmitter
Contracts, radius, or Pipe Primary Element   with Integral   Other Local   Temperature Element
Taps Connected to   Transmitter   Thermometer
Differential - Pressure   Temperature Indicator
Flow Transmitter

Venturi Tube   Level Indicator   Counting Switch,   Vibrance Transmitter   Weight Transmitter   Roll-Thickness
                          Float Type   Photoelectric, with   for Motor   Direct Connection   Transmitter
                                               Switch Action Based
                                               on Cumulative Total

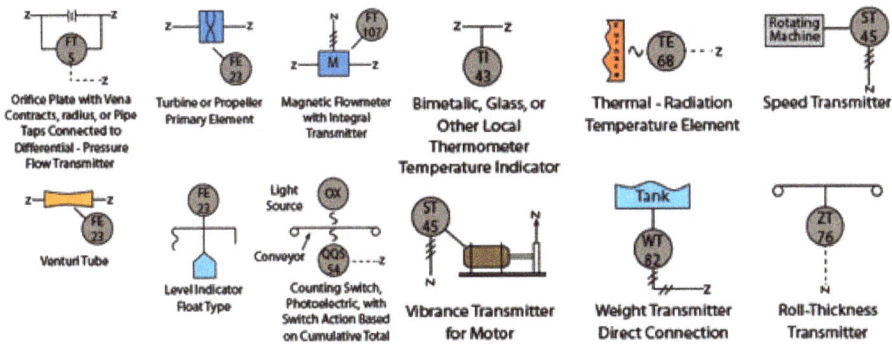

Primary elements

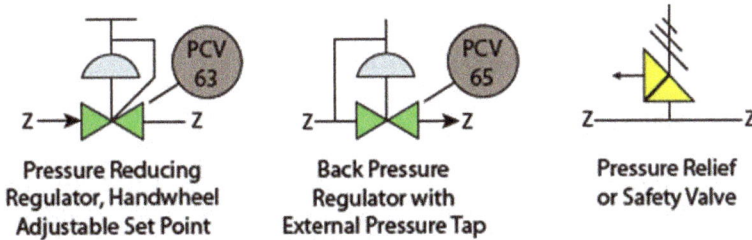

Pressure Reducing   Back Pressure   Pressure Relief
Regulator, Handwheel   Regulator with   or Safety Valve
Adjustable Set Point   External Pressure Tap

Regulators and safety valves

$A = \sqrt{X}$ $A = XY$ $A = X/Y$ $A = X-Y$

Math function

An Example of P&ID of a Small Process plant

Following is an example of P&ID of a mixing station:

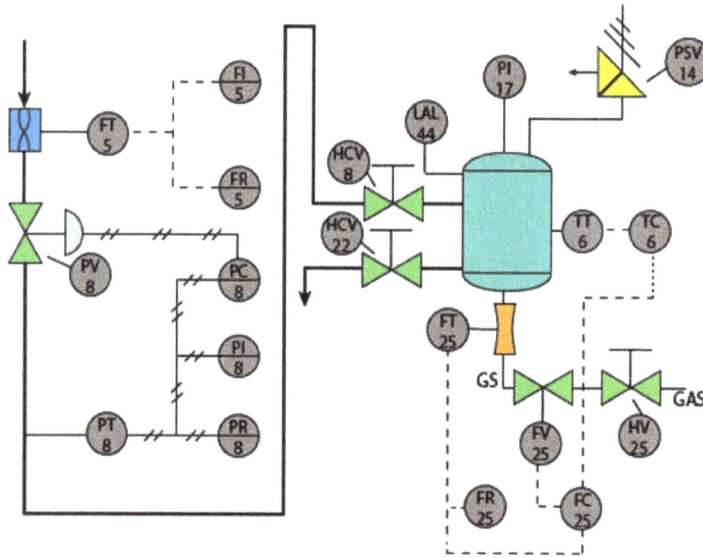

P&ID of a mixing station

A P&ID should not include the following:

- Instrument root valves
- Control relays
- Manual switches
- Equipment rating or capacity
- Primary instrument tubing and valves
- Pressure temperature and flow data
- Elbow, tees and similar standard fittings
- Extensive explanatory notes

A piping and instrumentation diagram/drawing (P&ID) is a detailed diagram in the process industry which shows the piping and vessels in the process flow, together with the instrumentation and control devices.

Superordinate to the *piping and instrumentation flowsheet* is the *process flow diagram (PFD)* which indicates the more general flow of plant processes and equipment and relationship between major equipment of a plant facility.

Contents and Function

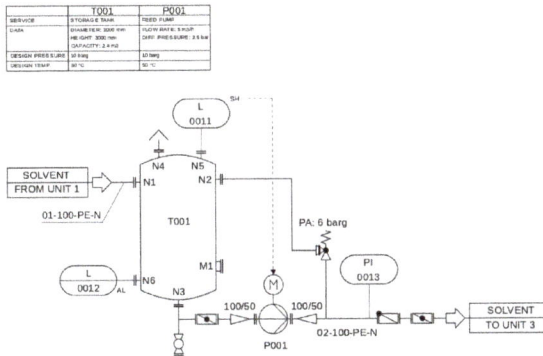

Piping and instrumentation diagram of pump with storage tank.
Symbols according to EN ISO 10628 and EN 62424.

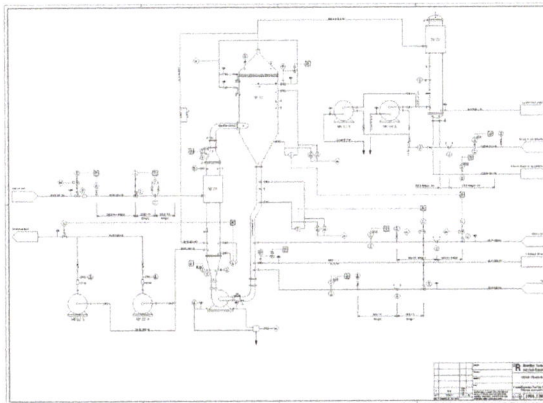

A more complex example of a P&ID.

A piping and instrumentation diagram/drawing (P&ID) is defined by the Institute of Instrumentation and Control as follows:

1. A diagram which shows the interconnection of process equipment and the instrumentation used to control the process. In the process industry, a standard set of symbols is used to prepare drawings of processes. The instrument symbols used in these drawings are generally based on International Society of Automation (ISA) Standard S5.1

2. The primary schematic drawing used for laying out a process control installation.

They usually contain the following information:

- Process piping, sizes and identification, including:

- o   Pipe classes or piping line numbers

- o   Flow directions

- o   Interconnections references

- o   Permanent start-up, flush and bypass lines

- Mechanical equipment/ Process control instrumentation and designation (names, numbers, unique tag identifiers), including:

  - o   Valves and their identifications (e.g. isolation, shutoff, relief and safety valves)

  - o   Control inputs and outputs (sensors and final elements, interlocks)

  - o   Miscellanea - vents, drains, flanges, special fittings, sampling lines, reducers and increasers

- Interfaces for class changes

- Computer control system

- Identification of components and subsystems delivered by others

P&IDs are originally drawn up at the design stage from a combination of process flow sheet data, the mechanical process equipment design, and the instrumentation engineering design. During the design stage, the diagram also provides the basis for the development of system control schemes, allowing for further safety and operational investigations, such as a Hazard and operability study (HAZOP). To do this, it is critical to demonstrate the physical sequence of equipment and systems, as well as how these systems connect.

P&IDs also play a significant role in the maintenance and modification of the process after initial build. Modifications are red-penned onto the diagrams and are vital records of the current plant design.

They are also vital in enabling development of;

- Control and shutdown schemes

- Safety and regulatory requirements

- Start-up sequences

- Operational understanding.

P&IDs form the basis for the live mimic diagrams displayed on graphical user interfaces of large industrial control systems such as SCADA and distributed control systems.

Identification and Reference Designation

Based on STANDARD ANSI/ISA S5.1 and ISO 14617-6, the P&ID is used for the identification of measurements within the process. The identifications consist of up to 5 letters. The first identification letter is for the measured value, the second is a modifier, 3rd indicates passive/readout function, 4th - active/output function, and the 5th is the function modifier. This is followed by loop number, which is unique to that loop. For instance FICO45 means it is the Flow Indicating Controller in control loop 045. This is also known as the "tag" identifier of the field device, which is normally given to the location and function of the instrument. The same loop may have FTO45 - which is the flow transmitter in the same loop.

Letter	Column 1 (Measured value)	Column 2 (Modifier)	Column 3 (Readout/passive function)	Column 4 (Output/active function)	Column 5 (Function modifier)
A	Analysis		Alarm		
B	Burner, combustion		User choice	User choice	User choice
C	User's choice (usually conductivity)			Control	Close
D	User's choice (usually density)	Difference			Deviation
E	Voltage		Sensor		
F	Flow rate	Ratio			
G	User's choice (usually gaging/ gauging)	Gas	Glass/gauge/viewing		
H	Hand				High
I	Current		Indicate		
J	Power	Scan			
K	Time, time schedule	Time rate of change		Control station	
L	Level		Light		Low
M	User's choice				Middle / intermediate
N	User's choice (usually torque)		User choice	User choice	User choice
O	User's choice		Orifice		Open
P	Pressure		Point/test connection		
Q	Quantity	Totalize/integrate	Totalize/integrate		
R	Radiation		Record		Run
S	Speed, frequency			Switch	Stop

T	Temperature			Transmit	
U	Multivariable		Multifunction	Multifunction	
V	Vibration, mechanical analysis			Valve or damper	
W	Weight, force		Well or probe		
X	User's choice (usually on-off valve as XV)	X-axis	Accessory devices, unclassified	Unclassified	Unclassified
Y	Event, state, presence	Y-axis		Auxiliary devices	
Z	Position, dimension	Z-axis or Safety		Actuator, driver or unclassified final control element	

For reference designation of any equipment in industrial systems the standard IEC 61346 (*Industrial systems, installations and equipment and industrial products — Structuring principles and reference designations*) can be applied. For the function *Measurement* the reference designator B is used, followed by the above listed letter for the measured variable.

For reference designation of any equipment in a power station the KKS Power Plant Classification System can be applied.

Symbols of Chemical Apparatus and Equipment

Below are listed some symbols of chemical apparatus and equipment normally used in a P&ID, according to ISO 10628 and ISO 14617.

Symbols of chemical apparatus and equipment									
	Pipe		Thermally insulated pipe		Jacketed pipe		Cooled or heated pipe		
	Jacketed mixing vessel (autoclave)		Half pipe mixing vessel		Pressurized horizontal vessel		Pressurized vertical vessel		
	Pump		Vacuum pump or compressor		Bag		Gas bottle		

	Fan		Axial fan, „MK,,,		Radial fan		Dryer
	Packed column		Tray column		Furnace		Cooling tower
	Heat ex- changer		Heat ex- changer		Cooler		Plate & frame heat ex- changer
	Double pipe heat exchang- er		Fixed straight tubes heat exchanger		U shaped tubes heat ex- changer		Spiral heat ex- changer
	Covered gas vent		Curved gas vent		(Air) filter		Funnel
	Steam trap		Viewing glass		Pressure reducing valve		Flexible pipe
	Valve		Control valve		Manual valve		Back draft damper
	Needle valve		Butterfly valve		Dia- phragm valve		Ball valve

Process Flow Diagram

A process flow diagram (PFD) is a diagram commonly used in chemical and process engineering to indicate the general flow of plant processes and equipment. The PFD displays the relationship between *major* equipment of a plant facility and does not show minor details such as piping details and designations. Another commonly used term for a PFD is a *flowsheet*.

Typical Content of a Process Flow Diagram

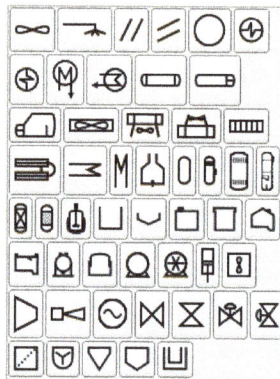

Some typical elements from process flow diagrams, as provided
by the open source program, Dia.

Typically, process flow diagrams of a single unit process will include the following:

- Process piping
- Major equipment items
- Control valves and other major valves
- Connections with other systems
- Major bypass and recirculation (recycle) streams
- Operational data (temperature, pressure, mass flow rate, density, etc.), often by stream references to a mass balance.
- Process stream names

Process flow diagrams generally do not include:

- Pipe classes or piping line numbers
- Process control instrumentation (sensors and final elements)
- Minor bypass lines
- Isolation and shutoff valves

- Maintenance vents and drains

- Relief and safety valves

- Flanges

Process flow diagrams of multiple process units within a large industrial plant will usually contain less detail and may be called *block flow diagrams* or *schematic flow diagrams*.

Process Flow Diagram Examples

The process flow diagram below depicts a single chemical engineering unit process known as an amine treating plant:

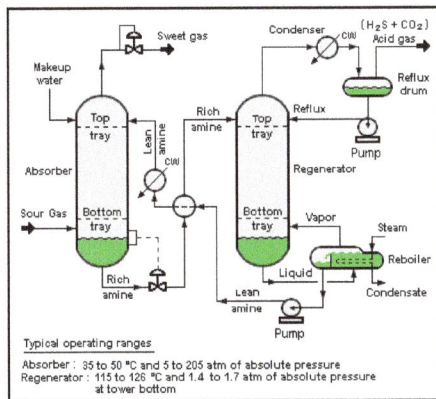

Flow diagram of a typical amine treating process used in industrial plants

Multiple Process Units within an Industrial Plant

The process flow diagram below is an example of a schematic or block flow diagram and depicts the various unit processes within a typical oil refinery:

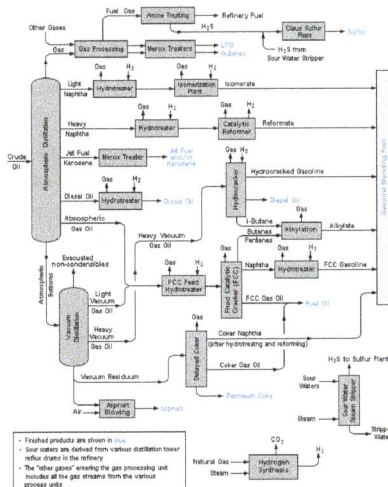

A typical oil refinery-SL

Other Items of Interest

A PFD can be computer generated from process simulators, CAD packages, or flow chart software using a library of chemical engineering symbols. Rules and symbols are available from standardization organizations such as DIN, ISO or ANSI. Often PFDs are produced on large sheets of paper.

PFDs of many commercial processes can be found in the literature, specifically in encyclopedias of chemical technology, although some might be outdated. To find recent ones, patent databases such as those available from the United States Patent and Trademark Office can be useful.

Standards

- ISO 10628: Flow Diagrams For Process Plants - General Rules

- ANSI Y32.11: Graphical Symbols For Process Flow Diagrams (withdrawn 2003)

- SAA AS 1109: Graphical Symbols For Process Flow Diagrams For The Food Industry

Process Flow Diagram, on the other hand, shows the relationships between the major components in the system. PFD also tabulates process design values for the components in different operating modes, typical minimum, normal and maximum. Following figure shows a typical PFD.

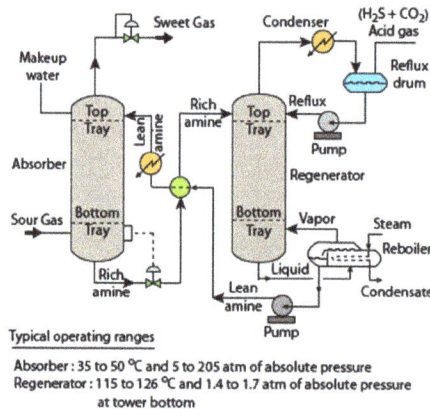

A typical process flow diagram

A PFD should include

- Major equipment symbols, names and identification numbers

- Control, valves and valves that affect operation of the system

- Interconnection with other systems

- Major bypass and recirculation lines

- System ratings and operational values as minimum, normal and maximum flow, temperature and pressure
- Composition of fluids

However it should never include

- Pipe class
- Pipe line numbers
- Minor bypass lines
- Isolation and shutoff valves
- Maintenance vents and drains
- Relief and safety valve
- Code class information
- Seismic class information

In other words, PFD has those information which are not contained in P&ID.

Actuators

Pneumatic Valve

Pneumatic valves are air-operated valves that are used for regulating fluid flow through a pipeline. Varying air pressure (3 to 15 psig) is used as an actuating signal for the pneumatic valves. Usually control signal is generated and transmitted in electrical form (4 to 20 mA) from the controller. Hence, a current to pressure converter (I/P) converts an analog signal (4-20 mA) to a proportional linear pneumatic output (3-15 psig). Its purpose is to translate the analog output from a controller into a precise, repeatable pressure value to control pneumatic actuators/operators, pneumatic valves, dampers, vanes, etc .

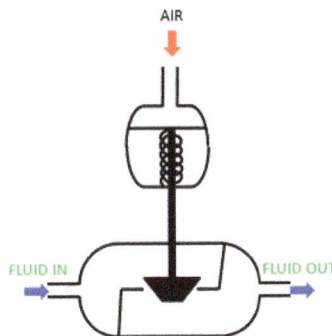

AIR

FLUID IN   FLUID OUT

(a) Fail Open

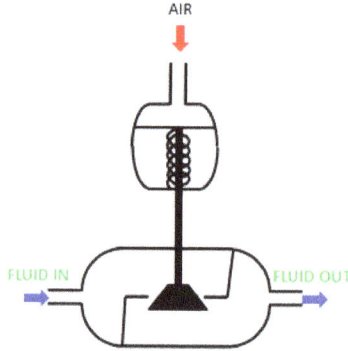

(b) Fail Closed
Pneumatics Valves

The bottom portion of the pneumatic valve has an orifice that separates the upstream and downstream flows. A tapered plug, capable of blocking the orifice in varying proportion, is attached to a stem that is connected to a diaphragm in the top portion of the valve. A spring restricts the movement of the stem. When air pressure above the diaphragm forces the stem to move downwards, the plug starts reducing the aperture of the orifice and eventually blocks the orifice at high air pressure. As a result the flow of fluid through the orifice is gradually decreased from "FULL FLOW" to "NO FLOW" condition. As the air pressure at the top of the diaphragm is released, the plug moves back to its original position resulting in full flow of fluid. This is called "FAIL OPEN" valve because when the control signal fails to provide enough air pressure, the valve remains in fully open condition. Similarly the "FAIL CLOSED" valve is shown in Figure (b) where the shape of the plug is opposite.

The dynamics of a typical pneumatic valve is usually second order in nature. The position of stem (or plug) determines the size of the aperture of the orifice that consequently determines the fluid flow rate. The position of the stem is determined by balancing all the forces acting on it. These forces are:

- Force exerted by the compressed air at the top of the diaphragm

- Force exerted by the spring

- Force exerted due to friction between stem and the valve packing

Hence,

$$pA - Kx - c\frac{dx}{dt} = M\frac{d^2x}{dt^2} \qquad (VII.1)$$

Where, p = pressure exerted by the compressed air, A = area of the diaphragm, x = displacement of stem, K = Hooke's constant of the spring, C = coefficient of friction between stem and the packing, M = mass of stem and its attachments.

Rearranging the eq. (VII.1), we obtain

$$\left(\frac{M}{K}\right)\frac{d^2x}{dt^2} + \left(\frac{C}{K}\right)\frac{dx}{dt} + x = \left(\frac{A}{K}\right)p \qquad (VII.2)$$

The transfer function between stem position and the actuating pressure is

$$\frac{\overline{x}(s)}{\overline{p}(s)} = \frac{\left(\dfrac{A}{K}\right)}{\left(\dfrac{M}{K}\right)s^2 + \left(\dfrac{C}{K}\right)s + 1} \qquad (VII.3)$$

which is indeed a second order dynamics. However, the movement of stem is always aided by proper lubrication which eventually reduces its coefficient of friction. Moreover, mass of the stem and its attachments are very small compared to the Hooke's constant ($M \ll K$ and $C \ll K$). Hence the dynamics of pneumatic valve can be neglected and movement of stem can be regarded as instantaneous in response to change of air pressure.

$$x = \left(\frac{A}{K}\right)p \qquad (VII.4)$$

Flow rate of non-flashing liquid is given by

$$F = C_v f(x)\sqrt{\frac{\Delta P}{\rho}} \qquad (VII.5)$$

Where, F = flow rate of liquid, ΔP = pressure drop across the upstream and downstream side of the valve, ρ = density of liquid, C_v = flow coefficient that depends on valve size, $f(x)$ = flow characteristic curve of the valve which is a function of the stem position x.

The shape of the flow characteristic curve depends on the shape of plug of the valve. There are five major types of flow characteristics used for designing the valve:

- Linear, $f(x) = x$

- Square root, $f(x) = \sqrt{x}$

- Equal percentage, $f(x) = \alpha^{x-1}$

- Hyperbolic, $f(x) = \{\alpha - (\alpha - 1)x\}^{-1}$

- Quick opening, $f(x) = x^{1/\alpha}$

Where, α = design constant. The types of plug are given in Figures.

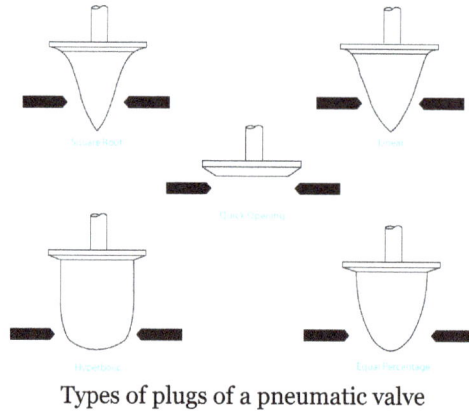

Types of plugs of a pneumatic valve

The typical flow characteristics curves are given in Figure.

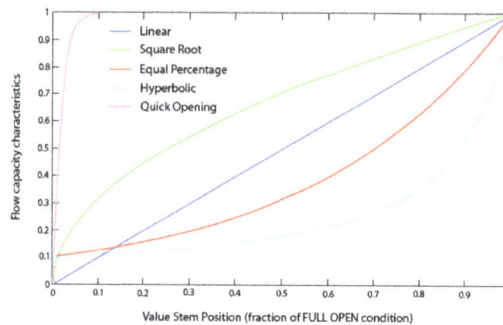

Flow capacity characteristics of various pneumatic valves

Pneumatic Actuator

A pneumatic control valve actuator converts energy (typically in the form of compressed air) into mechanical motion. The motion can be rotary or linear, depending on the type of actuator.

Principle of Operation

Globe control valve with pneumatic diaphragm actuator and "smart" positioner which will also feed back to the controller the actual valve position

Pneumatic rack and pinion actuators for valve controls of water pipes

A Pneumatic actuator mainly consists of a piston or a diaphragm which develops the motive power. It keeps the air in the upper portion of the cylinder, allowing air pressure to force the diaphragm or piston to move the valve stem or rotate the valve control element.

Valves require little pressure to operate and usually double or *triple* the input force. The larger the size of the piston, the larger the output pressure can be. Having a larger piston can also be good if air supply is low, allowing the same forces with less input. These pressures are large enough to crush objects in the pipe. On 100 kPa input, you could lift a small car (upwards of 1,000 lbs) easily, and this is only a basic, small pneumatic valve. However, the resulting forces required of the stem would be too great and cause the valve stem to fail.

This pressure is transferred to the valve stem, which is connected to either the valve plug, butterfly valve etc. Larger forces are required in high pressure or high flow pipelines to allow the valve to overcome these forces, and allow it to move the valves moving parts to control the material flowing inside.

The valves input is the "control signal." This can come from a variety of measuring devices, and each different pressure is a different set point for a valve. A typical standard signal is 20–100 kPa. For example, a valve could be controlling the pressure in a vessel which has a constant out-flow, and a varied in-flow (varied by the actuator and valve). A pressure transmitter will monitor the pressure in the vessel and transmit a signal from 20–100 kPa. 20 kPa means there is no pressure, 100 kPa means there is full range pressure (can be varied by the transmitters calibration points). As the pressure rises in the vessel, the output of the transmitter rises, this increase in pressure is sent to the valve, which causes the valve to stroke downward, and start closing the valve, decreasing flow into the vessel, reducing the pressure in the vessel as excess pressure is evacuated through the out flow. This is called a direct acting process.

Types

Some types of pneumatic actuators include:

- Tie rod cylinders

- Rotary actuators

- Grippers

- Rodless actuators with magnetic linkage or rotary cylinders

- Rodless actuators with mechanical linkage

- Pneumatic artificial muscles

- Speciality actuators that combine rotary and linear motion—frequently used for clamping operations

- Vacuum generators

Hydraulic Actuator

Hydraulic actuators employ hydraulic pressure to move a target device. These are used where high speed and large forces are required. Pressure applied to a confined incompressible fluid at any point is transmitted throughout the fluid in all directions and acts upon every part of the confining vessel at its interior surfaces. Figure represents the schematic of a hydraulic actuator.

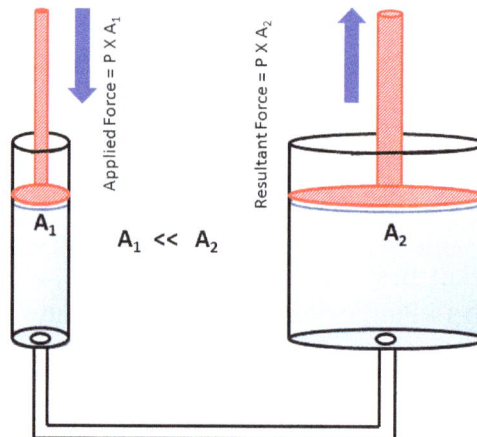

Hydraulic Actuator

According to Pascal's Law, since pressure P applied on an area A yields a force F, given as, $F = P \times A$, if a force is applied over a small area to cause a pressure P in a confined fluid, the force generated on a larger area can be made many times larger than the applied force that created the pressure.

Electric Actuator

Electric actuators can be of the following forms: Electric current/voltage, Solenoid, Stepping Motor and DC/AC Motor.

Current or voltage can easily be regulated to adjust the flow of energy into the process, e.g. heater, fan speed regulator etc. Energy supplied by the heater element is $W = I^2 Rt$, where I = current, R = resistance, t = time of heating. The current/voltage can be regulated using potentiometer (or rheostat), amplifier or a switch.

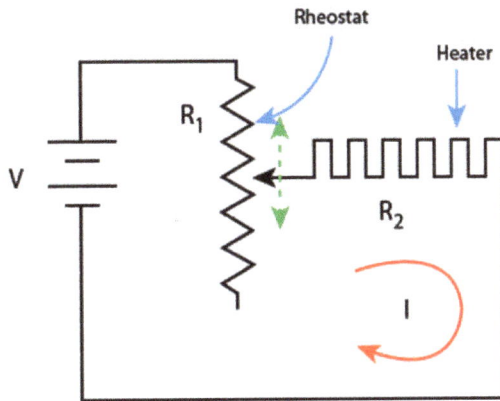

A schematic of Rheostat

A rheostat is a device that has variable resistance to current flow. The current flowing through the circuit is $I = V / (R_1 + R_2)$, where   can be varied by suitably sliding the pointer. The power transmitted to the heater would be $P = I^2 R_2$. As the pointer slides towards a lesser value of R_1, heater receives more power for heating

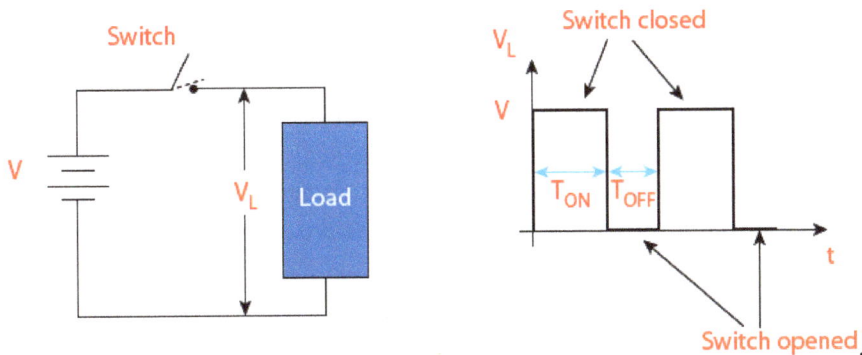

A schematic of Switch

A switch is a device which has two states of operation viz ., ON and OFF. The duration of the switch to remain in either state can be modulated as per requirement. If T_{ON} and T_{OFF} are the switch on and switch off time respectively, then duty cycle of the switch is defined as $T_{ON} / (T_{ON} + T_{OFF}) \times 100\%$ where $(T_{ON} + T_{OFF})$ is kept constant for any operation. This methods is often called as Pulse Width Modulation (PWM). Transistor and Thyristor are the examples of switches.

A solenoid is a coil wound into a tightly packed helix which produces a controlled magnetic field when an electric current is passed through it. Solenoids can be used as electromagnets which convert electromagnetic energy into linear motion of some mechanical part. It is often used as a valve which actuates a piston to restrict a flow in a pneumatic/hydraulic pipeline.

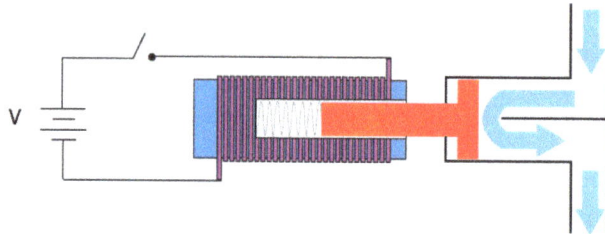

A schematic of Solenoid

Figure shows a schematic of a solenoid valve. The blue bar indicates an electromagnet that gets energized on flowing current through the helix. The red colored piston is a metallic object whose movement is controlled by the duration of energization of electromagnet. In normal situation the spring forces the piston to move far and block the fluid flow in the pipeline. As the electromagnet gets energized, the piston is pulled back yielding the fluid to flow unrestricted.

Sensors

Few types of such sensors will be discussed in the following subsections.

Temperature Measuring Devices

Most common devices for relatively low temperature measurement are thermocouples and resistance thermometers. Radiation pyrometers are used for high temperature measurements. Other temperature measurement devices are filled system thermometers, bimetal thermometers, oscillating quartz crystals, etc .

Thermocouple

A thermocouple is an electrical device consisting of two dissimilar conductors forming electrical junctions at differing temperatures. A thermocouple produces a temperature-dependent voltage as a result of the thermoelectric effect, and this voltage can be interpreted to measure temperature. Thermocouples are a widely used type of temperature sensor.

Commercial thermocouples are inexpensive, interchangeable, are supplied with standard connectors, and can measure a wide range of temperatures. In contrast to most other methods of temperature measurement, thermocouples are self powered and require no external form of excitation. The main limitation with thermocouples is accuracy; system errors of less than one degree Celsius (°C) can be difficult to achieve.

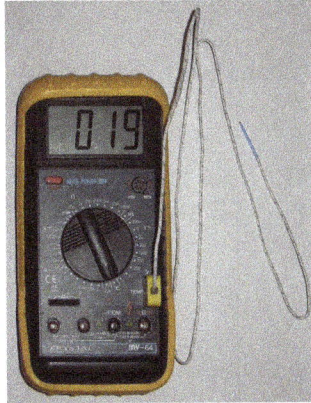

Thermocouple connected to a multimeter displaying room temperature in °C

Thermocouples are widely used in science and industry. Applications include temperature measurement for kilns, gas turbine exhaust, diesel engines, and other industrial processes. Thermocouples are also used in homes, offices and businesses as the temperature sensors in thermostats, and also as flame sensors in safety devices for gas-powered major appliances.

Principle of Operation

In 1821, the German physicist Thomas Johann Seebeck discovered that when different metals are joined at the ends and there is a temperature difference between the joints, a magnetic field is observed. At the time Seebeck referred to this as thermo-magnetism. The magnetic field he observed was later shown to be due to thermo-electric current. In practical use, the voltage generated at a single junction of two different types of wire is what is of interest as this can be used to measure temperature at very high and low temperatures. The magnitude of the voltage depends on the types of wire used. Generally, the voltage is in the microvolt range and care must be taken to obtain a usable measurement. Although current flows very little, power can be generated by a single thermocouple junction. Power generation using multiple thermocouples, as in a thermopile, is common.

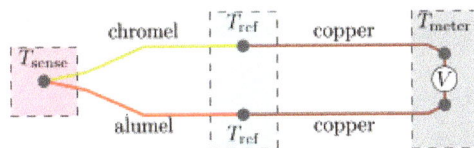

K-type thermocouple (chromel–alumel) in the standard thermocouple measurement configuration. The measured voltage V can be used to calculate temperature T_{sense}, provided that temperature T_{ref} is known.

The standard configuration for thermocouple usage is shown in the figure. Briefly, the desired temperature T_{sense} is obtained using three inputs—the characteristic function $E(T)$ of the thermocouple, the measured voltage V, and the reference junctions' temperature T_{ref}. The solution to the equation $E(T_{sense}) = V + E(T_{ref})$ yields T_{sense}. These details are often hidden from the user since the reference junction block (with T_{ref} thermometer), voltmeter, and equation solver are combined into a single product.

Physical Principle: Seebeck Effect

The Seebeck effect refers to an electromotive force whenever there is a temperature gradient in a conductive material. Under open-circuit conditions where there is no internal current flow, the gradient of voltage (∇V) is directly proportional to the gradient in temperature (∇T):

$$\nabla V = -S(T)\nabla T,$$

where $S(T)$ is a temperature-dependent material property known as the Seebeck coefficient.

The standard measurement configuration shown in the figure, shows four temperature regions and thus four voltage contributions:

1. Change from T_{meter} to T_{ref}, in the lower copper wire.

2. Change from T_{ref} to T_{sense}, in the alumel wire.

3. Change from T_{sense} to T_{ref}, in the chromel wire.

4. Change from T_{ref} to T_{meter}, in the upper copper wire.

The first and fourth contributions cancel out exactly, because these regions involve the same temperature change and an identical material. As a result, T_{meter} does not influence the measured voltage. The second and third contributions do not cancel, as they involve different materials.

The measured voltage turns out to be

$$V = \int_{T_{ref}}^{T_{sense}} \left(S_+(T) - S_-(T) \right) dT,$$

where S_+ and S_- are the Seebeck coefficients of the conductors attached to the positive and negative terminals of the voltmeter, respectively (chromel and alumel in the figure).

Characteristic Function

An integral does not need to be performed for every temperature measurement. Rather, the thermocouple's behaviour is captured by a characteristic function $E(T)$, which needs only to be consulted at two arguments:

$$V = E(T_{sense}) - E(T_{ref}).$$

In terms of the Seebeck coefficients, the characteristic function is defined by

$$E(T) = \int^{T} S_{+}(T') - S_{-}(T')dT' + const$$

The constant of integration in this indefinite integral has no significance, but is conventionally chosen such that $E(0°C) = 0$.

Thermocouple manufacturers and metrology standards organizations such as NIST provide tables of the function $E(T)$ that have been measured and interpolated over a range of temperatures, for particular thermocouple types.

Requirement for a Reference Junction

Reference junction block inside a Fluke CNX t3000 temperature meter. Two white wires connect to a thermistor (embedded in white thermal compound) to measure the reference junctions' temperature.

To obtain the desired measurement of T_{sense}, it is not sufficient to just measure V. The temperature at the reference junctions T_{ref} must be already known. Two strategies are often used here:

- "Ice bath" method: The reference junction block is immersed in a semi-frozen bath of distilled water at atmospheric pressure. The precise temperature of the melting point phase transition acts as a natural thermostat, fixing T_{ref} to 0 °C.

- Reference junction sensor (known as "cold junction compensation"): The reference junction block is allowed to vary in temperature, but the temperature is measured at this block using a separate temperature sensor. This secondary measurement is used to compensate for temperature variation at the junction block. The thermocouple junction is often exposed to extreme environments, while the reference junction is often mounted near the instrument's location. Semiconductor thermometer devices are often used in modern thermocouple instruments.

In both cases the value $V + E(T_{ref})$ is calculated, then the function $E(T)$ is searched for a matching value. The argument where this match occurs is the value of T_{sense}.

- A less common strategy is to use a hot reference junction. Historically, this was commonly found in temperature-critical processing plants where large numbers, often in the hundreds, of thermocouples were installed. It permitted the wiring from the field to the instrumentation or control room to be done using copper cable. Temperature control of the hot reference was either by an electrically heated, precision RTD controlled system or occasionally by a bimetallic controlled steam heater (in hazardous areas).

Practical Concerns

Thermocouples ideally should be very simple measurement devices, with each type being characterized by a precise $E(T)$ curve, independent of any other details. In reality, thermocouples are affected by issues such as alloy manufacturing uncertainties, aging effects, and circuit design mistakes/misunderstandings.

Circuit Construction

A common error in thermocouple construction is related to cold junction compensation. If an error is made on the estimation of T_{ref}, the same error will be carried over to the temperature measurement. For the simplest measurements, thermocouple wires are connected to copper far away from the hot or cold point whose temperature is measured; the cold junction is then assumed to be, at room temperature, but that temperature can vary.

Junctions should be made in a reliable manner, but there are many possible approaches to accomplish this. For low temperatures, junctions can be brazed or soldered, however it may be difficult to find a suitable flux and this may not be suitable at the sensing junction due to the solder's low melting point. Reference and extension junctions are therefore usually made with screw terminal blocks. For high temperatures, a common approach is a spot weld or crimp using a durable material. A common myth regarding thermocouples is that junctions must be made cleanly without involving a third metal, to avoid unwanted added emfs. This may result from another common misunderstanding that the voltage is generated at the junction. In fact, the junctions should in principle have uniform internal temperature, therefore no voltage is generated at the junction. The voltage is generated in the thermal gradient, along the wire.

A thermocouple produces small signals, often microvolts in magnitude. Precise measurements of this signal require an amplifier with low input offset voltage and with care taken to avoid thermal emfs from self-heating within the voltmeter itself. If the thermocouple wire has a high resistance for some reason (poor contact at junctions, or very thin wires used for fast thermal response), the measuring instrument should have high input impedance to prevent an offset in the measured voltage. A useful feature in thermocouple instrumentation will simultaneously measure resistance and detect faulty connections in the wiring or at thermocouple junctions.

Metallurgical Grades

While a thermocouple wire type is often described by its chemical composition, the actual aim is to produce a pair of wires that follow a standardized $E(T)$ curve.

Impurities affect each batch of metal differently, producing variable Seebeck coefficients. To match the standard behaviour, thermocouple wire manufacturers will deliberately mix in additional impurities to "dope" the alloy, compensating for uncontrolled variations in source material. As a result, there are standard and specialized grades of thermocouple wire, depending on the level of precision demanded in the thermocouple behaviour. Precision grades may only be available in matched pairs, where one wire is modified to compensate for deficiencies in the other wire.

A special case of thermocouple wire is known as "extension grade", designed to carry the thermoelectric circuit over a longer distance. Extension wires follow the stated $E(T)$ curve but for various reasons they are not designed to be used in extreme environments and so they cannot be used at the sensing junction in some applications. For example, an extension wire may be in a different form, such as highly flexible with stranded construction and plastic insulation, or be part of a multi-wire cable for carrying many thermocouple circuits. With expensive noble metal thermocouples, the extension wires may even be made of a completely different, cheaper material that mimics the standard type over a reduced temperature range.

Aging of Thermocouples

Thermocouples are often used at high temperatures and in reactive furnace atmospheres. In this case, the practical lifetime is limited by thermocouple aging. The thermoelectric coefficients of the wires in a thermocouple that is used to measure very high temperatures may change with time, and the measurement voltage accordingly drops. The simple relationship between the temperature difference of the junctions and the measurement voltage is only correct if each wire is homogeneous (uniform in composition). As thermocouples age in a process, their conductors can lose homogeneity due to chemical and metallurgical changes caused by extreme or prolonged exposure to high temperatures. If the aged section of the thermocouple circuit is exposed to a temperature gradient, the measured voltage will differ, resulting in error.

Aged thermocouples are only partly modified, for example being unaffected in the parts outside the furnace. For this reason, aged thermocouples cannot be taken out of their installed location and recalibrated in a bath or test furnace to determine error. This also explains why error can sometimes be observed when an aged thermocouple is pulled partly out of a furnace—as the sensor is pulled back, aged sections may see exposure to increased temperature gradients from hot to cold as the aged section now passes through the cooler refractory area, contributing significant error to the measurement. Likewise, an aged thermocouple that is pushed deeper into the furnace might sometimes provide a more accurate reading if being pushed further into the furnace causes the temperature gradient to occur only in a fresh section.

Types

Certain combinations of alloys have become popular as industry standards. Selection of the combination is driven by cost, availability, convenience, melting point, chemical properties, stability, and output. Different types are best suited for different applications. They are usually selected on the basis of the temperature range and sensitivity needed. Thermocouples with low sensitivities (B, R, and S types) have correspondingly lower resolutions. Other selection criteria include the chemical inertness of the thermocouple material and whether it is magnetic or not. Standard thermocouple types are listed below with the positive electrode (assuming $T_{sense} > T_{ref}$) first, followed by the negative electrode.

Nickel-alloy Thermocouples

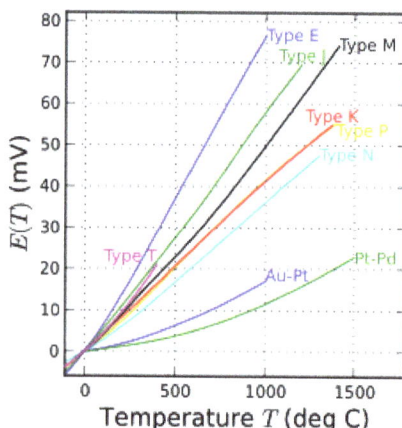

Characteristic functions for thermocouples that reach intermediate temperatures, as covered by nickel-alloy thermocouple types E, J, K, M, N, T. Also shown are the noble-metal alloy type P and the pure noble-metal combinations gold–platinum and platinum–palladium.

Type E

Type E (chromel–constantan) has a high output (68 μV/°C), which makes it well suited to cryogenic use. Additionally, it is non-magnetic. Wide range is −50 °C to +740 °C and narrow range is −110 °C to +140 °C.

Type J

Type J (iron–constantan) has a more restricted range (−40 °C to +750 °C) than type K but higher sensitivity of about 50 μV/°C. The Curie point of the iron (770 °C) causes a smooth change in the characteristic, which determines the upper temperature limit.

Type K

Type K (chromel–alumel) is the most common general-purpose thermocouple with a sensitivity of approximately 41 µV/°C. It is inexpensive, and a wide variety of probes are available in its −200 °C to +1350 °C (−330 °F to +2460 °F) range. Type K was specified at a time when metallurgy was less advanced than it is today, and consequently characteristics may vary considerably between samples. One of the constituent metals, nickel, is magnetic; a characteristic of thermocouples made with magnetic material is that they undergo a deviation in output when the material reaches its Curie point, which occurs for type K thermocouples at around 185 °C.

They operate very well in oxidizing atmospheres. If, however, a mostly reducing atmosphere (such as hydrogen with a small amount of oxygen) comes into contact with the wires, the chromium in the chromel alloy oxidizes. This reduces the emf output, and the thermocouple reads low. This phenomenon is known as *green rot*, due to the color of the affected alloy. Although not always distinctively green, the chromel wire will develop a mottled silvery skin and become magnetic. An easy way to check for this problem is to see whether the two wires are magnetic (normally, chromel is non-magnetic).

Hydrogen in the atmosphere is the usual cause of green rot. At high temperatures, it can diffuse through solid metals or an intact metal thermowell. Even a sheath of magnesium oxide insulating the thermocouple will not keep the hydrogen out.

Type M

Type M (82%Ni/18%Mo–99.2%Ni/0.8%Co, by weight) are used in vacuum furnaces for the same reasons as with type C (described below). Upper temperature is limited to 1400 °C. It is less commonly used than other types.

Type N

Type N (Nicrosil–Nisil) thermocouples are suitable for use between −270 °C and +1300 °C, owing to its stability and oxidation resistance. Sensitivity is about 39 µV/°C at 900 °C, slightly lower compared to type K.

Designed at the Defence Science and Technology Organisation (DSTO) of Australia, by Noel A. Burley, type-N thermocouples overcome the three principal characteristic types and causes of thermoelectric instability in the standard base-metal thermoelement materials:

1. A gradual and generally cumulative drift in thermal EMF on long exposure at elevated temperatures. This is observed in all base-metal thermoelement materials and is mainly due to compositional changes caused by oxidation, carburization, or neutron irradiation that can produce transmutation in nuclear reactor environments. In the case of type-K thermocouples, manganese and

aluminium atoms from the KN (negative) wire migrate to the KP (positive) wire, resulting in a down-scale drift due to chemical contamination. This effect is cumulative and irreversible.

2. A short-term cyclic change in thermal EMF on heating in the temperature range about 250–650 °C, which occurs in thermocouples of types K, J, T, and E. This kind of EMF instability is associated with structural changes such as magnetic short-range order in the metallurgical composition.

3. A time-independent perturbation in thermal EMF in specific temperature ranges. This is due to composition-dependent magnetic transformations that perturb the thermal EMFs in type-K thermocouples in the range about 25–225 °C, and in type J above 730 °C.

The Nicrosil and Nisil thermocouple alloys show greatly enhanced thermoelectric stability relative to the other standard base-metal thermocouple alloys because their compositions substantially reduce the thermoelectric instabilities described above. This is achieved primarily by increasing component solute concentrations (chromium and silicon) in a base of nickel above those required to cause a transition from internal to external modes of oxidation, and by selecting solutes (silicon and magnesium) that preferentially oxidize to form a diffusion-barrier, and hence oxidation-inhibiting films.

Type T

Type T (copper–constantan) thermocouples are suited for measurements in the −200 to 350 °C range. Often used as a differential measurement, since only copper wire touches the probes. Since both conductors are non-magnetic, there is no Curie point and thus no abrupt change in characteristics. Type-T thermocouples have a sensitivity of about 43 µV/°C. Note that copper has a much higher thermal conductivity than the alloys generally used in thermocouple constructions, and so it is necessary to exercise extra care with thermally anchoring type-T thermocouples.

Platinum/Rhodium-Alloy Thermocouples

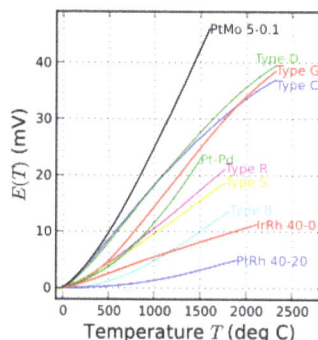

Characteristic functions for high-temperature thermocouple types, showing Pt/Rh, W/Re, Pt/Mo, and Ir/Rh-alloy thermocouples. Also shown is the Pt–Pd pure-metal thermocouple.

Types B, R, and S thermocouples use platinum or a platinum/rhodium alloy for each conductor. These are among the most stable thermocouples, but have lower sensitivity than other types, approximately 10 µV/°C. Type B, R, and S thermocouples are usually used only for high-temperature measurements due to their high cost and low sensitivity.

Type B

Type B (70%Pt/30%Rh−94%Pt/6%Rh, by weight) thermocouples are suited for use at up to 1800 °C. Type-B thermocouples produce the same output at 0 °C and 42 °C, limiting their use below about 50 °C. The emf function has a minimum around 21 °C, meaning that cold-junction compensation is easily performed, since the compensation voltage is essentially a constant for a reference at typical room temperatures.

Type R

Type R (87%Pt/13%Rh−Pt, by weight) thermocouples are used up to 1600 °C.

Type S

Type S (90%Pt/10%Rh−Pt, by weight) thermocouples, similar to type R, are used up to 1600 °C. Before the introduction of the International Temperature Scale of 1990 (ITS-90), precision type-S thermocouples were used as the practical standard thermometers for the range of 630 °C to 1064 °C, based on an interpolation between the freezing points of antimony, silver, and gold. Starting with ITS-90, platinum resistance thermometers have taken over this range as standard thermometers.

Tungsten/Rhenium-alloy Thermocouples

These thermocouples are well suited for measuring extremely high temperatures. Typical uses are hydrogen and inert atmospheres, as well as vacuum furnaces. They are not used in oxidizing environments at high temperatures because of embrittlement. A typical range is 0 to 2315 °C, which can be extended to 2760 °C in inert atmosphere and to 3000 °C for brief measurements.

Type C

(95%W/5%Re−74%W/26%Re, by weight)

Type D

(97%W/3%Re−75%W/25%Re, by weight)

Type G

(W−74%W/26%Re, by weight)

Others

Chromel–Gold/Iron-alloy Thermocouples

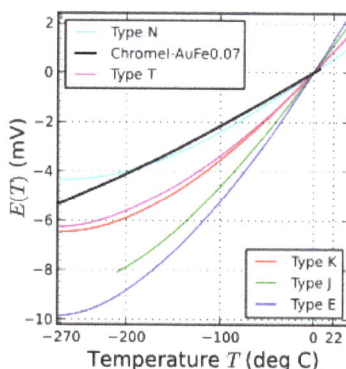

Thermocouple characteristics at low temperatures. The AuFe-based thermocouple shows a steady sensitivity down to low temperatures, whereas conventional types soon flatten out and lose sensitivity at low temperature.

In these thermocouples (chromel–gold/iron alloy), the negative wire is gold with a small fraction (0.03–0.15 atom percent) of iron. The impure gold wire gives the thermocouple a high sensitivity at low temperatures (compared to other thermocouples at that temperature), whereas the chromel wire maintains the sensitivity near room temperature. It can be used for cryogenic applications (1.2–300 K and even up to 600 K). Both the sensitivity and the temperature range depend on the iron concentration. The sensitivity is typically around 15 μV/K at low temperatures, and the lowest usable temperature varies between 1.2 and 4.2 K.

Type P (Noble-Metal Alloy)

Type P (55%Pd/31%Pt/14%Au–65%Au/35%Pd, by weight) thermocouples give a thermoelectric voltage that mimics the type K over the range 500 °C to 1400 °C, however they are constructed purely of noble metals and so shows enhanced corrosion resistance. This combination is also known as Platinel II.

Platinum/Molybdenum-Alloy Thermocouples

Thermocouples of platinum/molybdenum-alloy (95%Pt/5%Mo–99.9%Pt/0.1%Mo, by weight) are sometimes used in nuclear reactors, since they show a low drift from nuclear transmutation induced by neutron irradiation, compared to the platinum/rhodium-alloy types.

Iridium/Rhodium Alloy Thermocouples

The use of two wires of iridium/rhodium alloys can provide a thermocouple that can be used up to about 2000 °C in inert atmospheres.

Pure Noble-metal Thermocouples Au–Pt, Pt–Pd

Thermocouples made from two different, high-purity noble metals can show high accuracy even when uncalibrated, as well as low levels of drift. Two combinations in use are gold–platinum and platinum–palladium. Their main limitations are the low melting points of the metals involved (1064 °C for gold and 1555 °C for palladium). These thermocouples tend to be more accurate than type S, and due to their economy and simplicity are even regarded as competitive alternatives to the platinum resistance thermometers that are normally used as standard thermometers.

Skutterudite Thermocouples

NASA is developing a Multi-Mission Radioisotope Thermoelectric Generator in which the thermocouples would be made of skutterudite, which can function with a smaller temperature difference than the current tellurium designs. This would mean that an otherwise similar RTG would generate 25% more power at the beginning of a mission and at least 50% more after seventeen years. NASA hopes to use the design on the next New Frontiers mission.

Comparison of Types

The table below describes properties of several different thermocouple types. Within the tolerance columns, T represents the temperature of the hot junction, in degrees Celsius. For example, a thermocouple with a tolerance of $\pm 0.0025 \times T$ would have a tolerance of ± 2.5 °C at 1000 °C.

Type	Temperature range (°C)				Tolerance class (°C)		Color code		
	Continuous		Short-term		One	Two	IEC	BS	ANSI
	Low	High	Low	High					
K	0	+1100	−180	+1300	$-40 - 375$: ± 1.5 $375 - 1000$: $\pm 0.004 \times T$	$-40 - 333$: ± 2.5 $333 - 1200$: $\pm 0.0075 \times T$			
J	0	+750	−180	+800	$-40 - 375$: ± 1.5 $375 - 750$: $\pm 0.004 \times T$	$-40 - 333$: ± 2.5 $333 - 750$: $\pm 0.0075 \times T$			
N	0	+1100	−270	+1300	$-40 - 375$: ± 1.5 $375 - 1000$: $\pm 0.004 \times T$	$-40 - 333$: ± 2.5 $333 - 1200$: $\pm 0.0075 \times T$			

Type					Tolerance class 1	Tolerance class 2			Polarity
R	0	+1600	−50	+1700	0 − 1100: ±1.0 1100 − 1600: ±0.003×(T−767)	0 − 600: ±1.5 600 − 1600: ±0.0025×T	+ / −	+ / −	Not defined
S	0	+1600	−50	+1750	0 − 1100: ±1.0 1100 − 1600: ±0.003×(T−767)	0 − 600: ±1.5 600 − 1600: ±0.0025×T		+ / −	Not defined
B	+200	+1700	0	+1820	Not available	600 − 1700: ±0.0025×T	No standard	No standard	Not defined
T	−185	+300	−250	+400	−40 − 125: ±0.5 125 − 350: ±0.004×T	−40 − 133: ±1.0 133 − 350: ±0.0075×T	+ / −	+ / −	+ / −
E	0	+800	−40	+900	−40 − 375: ±1.5 375 − 800: ±0.004×T	−40 − 333: ±2.5 333 − 900: ±0.0075×T	+ / −	+ / −	+ / −
Chromel/AuFe	−272	+300	N/A	N/A	Reproducibility 0.2% of the voltage. Each sensor needs individual calibration.				

Thermocouple Insulation

Typical low cost type K thermocouple (with standard type K connector). While the wires can survive and function at high temperatures, the plastic insulation will start to break down at 300 °C.

The wires that make up the thermocouple must be insulated from each other everywhere, except at the sensing junction. Any additional electrical contact between the wires, or contact of a wire to other conductive objects, can modify the voltage and give a false reading of temperature.

Plastics are suitable insulators for low temperatures parts of a thermocouple, whereas ceramic insulation can be used up to around 1000 °C. Other concerns (abrasion and chemical resistance) also affect the suitability of materials.

When wire insulation disintegrates, it can result in an unintended electrical contact at a different location from the desired sensing point. If such a damaged thermocouple is used in the closed loop control of a thermostat or other temperature controller, this can lead to a runaway overheating event and possibly severe damage, as the false temperature reading will typically be lower than the sensing junction temperature. Failed insulation will also typically outgas, which can lead to process contamination. For parts of thermocouples used at very high temperatures or in contamination-sensitive applications, the only suitable insulation may be vacuum or inert gas; the mechanical rigidity of the thermocouple wires is used to keep them separated.

Table of Insulation Materials

Type of Insulation	Max. continuous temperature	Max. single reading	Abrasion resistance	Moisture resistance	Chemical resistance
Mica–glass tape	649 °C/1200 °F	705 °C/1300 °F	Good	Fair	Good
TFE tape, TFE–glass tape	649 °C/1200 °F	705 °C/1300 °F	Good	Fair	Good
Vitreous-silica braid	871 °C/1600 °F	1093 °C/2000 °F	Fair	Poor	Poor
Double glass braid	482 °C/900 °F	538 °C/1000 °F	Good	Good	Good
Enamel–glass braid	482 °C /900 °F	538 °C/1000 °F	Fair	Good	Good
Double glass wrap	482 °C/900 °F	427 °C/800 °F	Fair	Good	Good
Non-impregnated glass braid	482 °C/900 °F	427 °C/800 °F	Poor	Poor	Fair
Skive TFE tape, TFE–glass braid	482 °C/900 °F	538 °C/1000 °F	Good	Excellent	Excellent
Double cotton braid	88 °C/190 °F	120 °C/248 °F	Good	Good	Poor
"S" glass with binder	704 °C/1300 °F	871 °C/1600 °F	Fair	Fair	Good
Nextel ceramic fiber	1204 °C/2200 °F	1427 °C/2600 °F	Fair	Fair	Fair
Polyvinyl/nylon	105 °C/221 °F	120 °C/248 °F	Excellent	Excellent	Good
Polyvinyl	105 °C/221 °F	105 °C/221 °F	Good	Excellent	Good
Nylon	150 °C/302 °F	130 °C/266 °F	Excellent	Good	Good
PVC	105 °C/221 °F	105 °C/221 °F	Good	Excellent	Good
FEP	204 °C/400 °F	260 °C/500 °F	Excellent	Excellent	Excellent
Wrapped and fused TFE	260 °C/500 °F	316 °C/600 °F	Good	Excellent	Excellent
Kapton	316 °C/600 °F	427 °C/800 °F	Excellent	Excellent	Excellent
Tefzel	150 °C/302 °F	200 °C/392 °F	Excellent	Excellent	Excellent
PFA	260 °C/500 °F	290 °C/550 °F	Excellent	Excellent	Excellent
T300*	300 °C	–	Good	Excellent	Excellent

Temperature ratings for insulations may vary based on what the overall thermocouple construction cable consists of.

Note: T300 is a new high-temperature material that was recently approved by UL for 300 °C operating temperatures.

Applications

Thermocouples are suitable for measuring over a large temperature range, from −270 up to 3000 °C (for a short time, in inert atmosphere). Applications include temperature measurement for kilns, gas turbine exhaust, diesel engines, other industrial processes and fog machines. They are less suitable for applications where smaller temperature differences need to be measured with high accuracy, for example the range 0–100 °C with 0.1 °C accuracy. For such applications thermistors, silicon bandgap temperature sensors and resistance thermometers are more suitable.

Steel Industry

Type B, S, R and K thermocouples are used extensively in the steel and iron industries to monitor temperatures and chemistry throughout the steel making process. Disposable, immersible, type S thermocouples are regularly used in the electric arc furnace process to accurately measure the temperature of steel before tapping. The cooling curve of a small steel sample can be analyzed and used to estimate the carbon content of molten steel.

Gas Appliance Safety

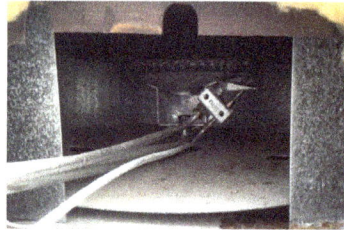

A thermocouple (the right most tube) inside the burner assembly of a water heater

Thermocouple connection in gas appliances. The end ball (contact) on the left is insulated from the fitting by an insulating washer. The thermocouple line consists of copper wire, insulator and outer metal (usually copper) sheath which is also used as ground.

Many gas-fed heating appliances such as ovens and water heaters make use of a pilot flame to ignite the main gas burner when required. If the pilot flame goes out, un-burned gas may be released, which is an explosion risk and a health hazard. To prevent this, some appliances use a thermocouple in a fail-safe circuit to sense when the pilot light is burning. The tip of the thermocouple is placed in the pilot flame, generating a voltage which operates the supply valve which feeds gas to the pilot. So long as the pilot flame remains lit, the thermocouple remains hot, and the pilot gas valve is held open. If the pilot light goes out, the thermocouple temperature falls, causing the voltage across the thermocouple to drop and the valve to close.

Some combined main burner and pilot gas valves (mainly by Honeywell) reduce the power demand to within the range of a single universal thermocouple heated by a pilot (25 mV open circuit falling by half with the coil connected to a 10–12 mV, 0.2–0.25 A source, typically) by sizing the coil to be able to hold the valve open against a light spring, but only after the initial turning-on force is provided by the user pressing and holding a knob to compress the spring during lighting of the pilot. These systems are identifiable by the "press and hold for x minutes" in the pilot lighting instructions. (The holding current requirement of such a valve is much less than a bigger solenoid de-signed for pulling the valve in from a closed position would require.) Special test sets are made to confirm the valve let-go and holding currents, because an ordinary mil-liammeter cannot be used as it introduces more resistance than the gas valve coil. Apart from testing the open circuit voltage of the thermocouple, and the near short-circuit DC continuity through the thermocouple gas valve coil, the easiest non-specialist test is substitution of a known good gas valve.

Some systems, known as millivolt control systems, extend the thermocouple concept to both open and close the main gas valve as well. Not only does the voltage created by the pilot thermocouple activate the pilot gas valve, it is also routed through a ther-mostat to power the main gas valve as well. Here, a larger voltage is needed than in a pilot flame safety system described above, and a thermopile is used rather than a single thermocouple. Such a system requires no external source of electricity for its operation and thus can operate during a power failure, provided that all the other related system components allow for this. This excludes common forced air furnaces because external electrical power is required to operate the blower motor, but this feature is especial-ly useful for un-powered convection heaters. A similar gas shut-off safety mechanism using a thermocouple is sometimes employed to ensure that the main burner ignites within a certain time period, shutting off the main burner gas supply valve should that not happen.

Out of concern about energy wasted by the standing pilot flame, designers of many newer appliances have switched to an electronically controlled pilot-less ignition, also called intermittent ignition. With no standing pilot flame, there is no risk of gas build-up should the flame go out, so these appliances do not need thermocouple-based pilot safety switches. As these designs lose the benefit of operation without a continuous source of electricity, standing pilots are still used in some appliances. The exception is later model instantaneous (aka "tankless") water heaters that use the flow of water to generate the current required to ignite the gas burner; these designs also use a thermocouple as a safety cut-off device in the event the gas fails to ignite, or if the flame is extinguished.

Thermopile Radiation Sensors

Thermopiles are used for measuring the intensity of incident radiation, typically visible or infrared light, which heats the hot junctions, while the cold junctions are on a heat sink. It is possible to measure radiative intensities of only a few $\mu W/cm^2$ with commercially available thermopile sensors. For example, some laser power meters are based on such sensors; these are specifically known as thermopile laser sensor.

The principle of operation of a thermopile sensor is distinct from that of a bolometer, as the latter relies on a change in resistance.

Manufacturing

Thermocouples can generally be used in the testing of prototype electrical and mechanical apparatus. For example, switchgear under test for its current carrying capacity may have thermocouples installed and monitored during a heat run test, to confirm that the temperature rise at rated current does not exceed designed limits.

Power Production

A thermocouple can produce current to drive some processes directly, without the need for extra circuitry and power sources. For example, the power from a thermocouple can activate a valve when a temperature difference arises. The electrical energy generated by a thermocouple is converted from the heat which must be supplied to the hot side to maintain the electric potential. A continuous transfer of heat is necessary because the current flowing through the thermocouple tends to cause the hot side to cool down and the cold side to heat up (the Peltier effect).

Thermocouples can be connected in series to form a thermopile, where all the hot junctions are exposed to a higher temperature and all the cold junctions to a lower temperature. The output is the sum of the voltages across the individual junctions, giving larger voltage and power output. In a radioisotope thermoelectric generator, the radioactive decay of transuranic elements as a heat source has been used to power spacecraft on missions too far from the Sun to use solar power.

Thermopiles heated by kerosene lamps were used to run batteryless radio receivers in isolated areas. There are commercially produced lanterns that use the heat from a candle to run several light-emitting diodes, and thermoelectrically-powered fans to improve air circulation and heat distribution in wood stoves.

Process plants

Chemical production and petroleum refineries will usually employ computers for logging and for limit testing the many temperatures associated with a process, typically numbering in the hundreds. For such cases, a number of thermocouple leads will be brought to a common reference block (a large block of copper) containing the second thermocouple of each circuit. The temperature of the block is in turn measured by a thermistor. Simple computations are used to determine the temperature at each measured location.

Thermocouple as Vacuum Gauge

A thermocouple can be used as a vacuum gauge over the range of approximately 0.001 to 1 torr absolute pressure. In this pressure range, the mean free path of the gas is comparable to the dimensions of the vacuum chamber, and the flow regime is neither purely viscous nor purely molecular. In this configuration, the thermocouple junction is attached to the centre of a short heating wire, which is usually energised by a constant current of about 5 mA, and the heat is removed at a rate related to the thermal conductivity of the gas. It may be possible to superimpose AC heating on the thermocouple directly, making the sensor a 2-wire device, but those on the market appear to all be 4-wire devices, with separate terminals for the heater and the thermocouple.

The temperature detected at the thermocouple junction depends on the thermal conductivity of the surrounding gas, which depends on the pressure of the gas. The potential difference measured by a thermocouple is proportional to the square of pressure over the low- to medium-vacuum range. At higher (viscous flow) and lower (molecular flow) pressures, the thermal conductivity of air or any other gas is essentially independent of pressure. The thermocouple was first used as a vacuum gauge by Voege in 1906. The mathematical model for the thermocouple as a vacuum gauge is quite complicated, as explained in detail by Van Atta, but can be simplified to:

$$P = \frac{B(V^2 - V_0^2)}{V_0^2},$$

where P is the gas pressure, B is a constant that depends on the thermocouple temperature, the gas composition and the vacuum-chamber geometry, V_0 is the thermocouple voltage at zero pressure (absolute), and V is the voltage indicated by the thermocouple.

The alternative is the Pirani gauge, which operates in a similar way, over approximately the same pressure range, but is only a 2-terminal device, sensing the change in resistance with temperature of a thin electrically heated wire, rather than using a thermocouple.

Thermoelectric Generator

A thermoelectric generator (TEG), also called a Seebeck generator, is a solid state device that converts heat flux (temperature differences) directly into electrical energy through a phenomenon called the Seebeck effect (a form of thermoelectric effect). Thermoelectric generators function like heat engines, but are less bulky and have no moving parts. However, TEGs are typically more expensive and less efficient.

Thermoelectric generators could be used in power plants in order to convert waste heat into additional electrical power and in automobiles as automotive thermoelectric generators (ATGs) to increase fuel efficiency. Another application is radioisotope thermoelectric generators which are used in space probes, which has the same mechanism but use radioisotopes to generate the required heat difference.

History

In 1821, Thomas Johann Seebeck discovered that a thermal gradient formed between two dissimilar conductors can produce electricity. At the heart of the thermoelectric effect is the fact that a temperature gradient in a conducting material results in heat flow; this results in the diffusion of charge carriers. The flow of charge carriers between the hot and cold regions in turn creates a voltage difference. In 1834, Jean Charles Athanase Peltier discovered the reverse effect, that running an electric current through the junction of two dissimilar conductors could, depending on the direction of the current, cause it to act as a heater or cooler.

Construction

Thermoelectric power generators consist of three major components: thermoelectric materials, thermoelectric modules and thermoelectric systems that interface with the heat source.

Thermoelectric Materials

Thermoelectric materials generate power directly from heat by converting temperature differences into electric voltage. These materials must have both high electrical conductivity (σ) and low thermal conductivity (κ) to be good thermoelectric materials. Having low thermal conductivity ensures that when one side is made hot, the other side stays cold, which helps to generate a large voltage while in a temperature gradient. The measure of the magnitude of electrons flow in response to a temperature difference across that material is given by the Seebeck coefficient (S). The efficiency of a given material to produce a thermoelectric power is governed by its "figure of merit" $zT = S^2\sigma T/\kappa$.

For many years, the main three semiconductors known to have both low thermal conductivity and high power factor were bismuth telluride (Bi_2Te_3), lead telluride (PbTe), and silicon germanium (SiGe). These materials have very rare elements which make them very expensive compounds.

Today, the thermal conductivity of semiconductors can be lowered without affecting their high electrical properties using nanotechnology. This can be achieved by creating nanoscale features such as particles, wires or interfaces in bulk semiconductor materials. However, the manufacturing processes of nano-materials is still challenging.

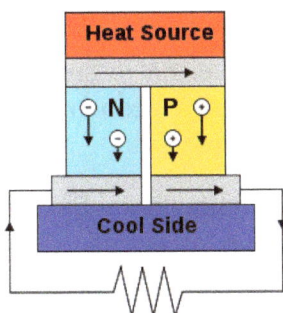

A thermoelectric circuit composed of materials of different Seebeck coefficient (p-doped and n-doped semiconductors), configured as a thermoelectric generator.

Thermoelectric Module

A thermoelectric module is a circuit containing thermoelectric materials that generate electricity from heat directly. A thermoelectric module consists of two dissimilar thermoelectric materials joining in their ends: an n-type (negatively charged); and a p-type (positively charged) semiconductors. A direct electric current will flow in the circuit when there is a temperature difference between the two materials. Generally, the current magnitude has a proportional relationship with the temperature difference. (i.e., the more the temperature difference, the higher the current.)

In application, thermoelectric modules in power generation work in very tough mechanical and thermal conditions. Because they operate in very high temperature gradient, the modules are subject to large thermally induced stresses and strains for long periods of time. They also are subject to mechanical fatigue caused by large number of thermal cycles.

Thus, the junctions and materials must be selected so that they survive these tough mechanical and thermal conditions. Also, the module must be designed such that the two thermoelectric materials are thermally in parallel, but electrically in series. The efficiency of thermoelectric modules are greatly affected by its geometrical design.

Thermoelectric System

Using thermoelectric modules, a thermoelectric system generates power by taking in heat from a source such as a hot exhaust flue. In order to do that, the system needs a large temperature gradient, which is not easy in real-world applications. The cold side must be cooled by air or water. Heat exchangers are used on both sides of the modules to supply this heating and cooling.

There are many challenges in designing a reliable TEG system that operates at high temperatures. Achieving high efficiency in the system requires extensive engineering design in order to balance between the heat flow through the modules and maximizing the temperature gradient across them. To do this, designing heat exchanger technologies in the system is one of the most important aspects of TEG engineering. In addition, the system requires to minimize the thermal losses due to the interfaces between materials at several places. Another challenging constraint is avoiding large pressure drops between the heating and cooling sources.

After the DC power from the TE modules passes through an inverter, the TEG produces AC power, which in turn, requires an integrated power electronics system to deliver it to the customer.

Materials for TEG

Only a few known materials to date are identified as thermoelectric materials. Most thermoelectric materials today have a ZT, the figure of merit, value of around unity, such as in Bismuth Telluride (Bi_2Te_3) at room temperature and lead telluride (PbTe) at 500-700K. However, in order to be competitive with other power generation systems, TEG materials should have zT of 2-3 range. Most research in thermoelectric materials has focused on increasing the Seebeck coefficient (S) and reducing the thermal conductivity, especially by manipulating the nanostructure of the thermoelectric materials. Because the thermal and electrical conductivity correlate with the charge carriers, new means must be introduced in order to conciliate the contradiction between high electrical conductivity and low thermal conductivity as indicated.

When selecting materials for thermoelectric generation, a number of other factors need to be considered. During operation, ideally the thermoelectric generator has a large temperature gradient across it. Thermal expansion will then introduce stress in the device which may cause fracture of the thermoelectric legs, or separation from the coupling material. The mechanical properties of the materials must be considered and the coefficient of thermal expansion of the n and p-type material must be matched reasonably well. In segmented thermoelectric generators, the material's compatibility must also be considered. A material's compatibility factor is defined as

$s = \left(\dfrac{\sqrt{1-zT}-1}{ST} \right)$. When the compatibility factor from one segment to the next differs by more than a factor of about two, the device will not operate efficiently. The material parameters determining s (as well as zT) are temperature dependent, so the compatibility factor may change from the hot side to the cold side of the device, even in one segment. This behavior is referred to as self-compatibility and may become important in devices design for low temperature operation.

In general, thermoelectric materials can be categorized into conventional and new materials:

Conventional Materials

There are many TEG materials that are employed in commercial applications today. These materials can be divided into three groups based on the temperature range of operation:

1. Low temperature materials (up to around 450K): Alloys based on Bismuth (Bi) in combinations with Antimony (Sb), Tellurium (Te) or Selenium (Se).

2. Intermediate temperature (up to 850K): such as materials based on alloys of Lead (Pb)

3. Highest temperatures material (up to 1300K): materials fabricated from silicon germanium (SiGe) alloys.

Although these materials still remain the cornerstone for commercial and practical applications in thermoelectric power generation, significant advances have been made in synthesizing new materials and fabricating material structures with improved thermoelectric performance. Recent research have focused on improving the material's figure-of-merit (zT), and hence the conversion efficiency, by reducing the lattice thermal conductivity.

New Materials

Researchers are trying to develop new thermoelectric materials for power generation by improving the figure-of-merit zT. One example of these materials is the semiconductor compound β-Zn_4Sb_3, which possesses an exceptionally low thermal conductivity and exhibits a maximum zT of 1.3 at a temperature of 670K. This material is also relatively inexpensive and stable up to this temperature in a vacuum, and can be a good alternative in the temperature range between materials based on Bi_2Te_3 and PbTe.

Beside improving the figure-of-merit, there is increasing focus to develop new materials by increasing the electrical power output, decreasing cost and developing environmentally friendly materials. For example, when the fuel cost is low or almost free, such as in waste heat recovery, then the cost per watt is only determined by the power per unit area and the operating period. As a result, it has initiated a search for materials with high power output rather than conversion efficiency. For example, the rare earth compounds $YbAl_3$ has a low figure-of-merit, but it has a power output of at least double that of any other material, and can operate over the temperature range of a waste heat source.

Novel Processing

In order to increase the figure of merit (zT), a material's thermal conductivity should be minimized while its electrical conductivity and Seebeck coefficient is maximized. In most cases, methods to increase or decrease one property result in the same effect on other properties due to their interdependence. A novel processing technique exploits the scattering of different phonon frequencies to selectively reduce lattice thermal conductivity without the typical negative effects on electrical conductivity from the simultaneous increased scattering of electrons. In a bismuth antimony tellurium ternary system, liquid-phase sintering is used to produce low-energy semicoherent grain boundaries, which do not have a significant scattering effect on electrons. The breakthrough is then applying a pressure to the liquid in the sintering process, which creates a transient flow of the Te rich liquid and facilitates the formation of dislocations that greatly reduce the lattice conductivity. The ability to selectively decrease the lattice conductivity results in reported zT values of 1.86 ± .15 which are a significant improvement over current commercial thermoelectric generators which have typical figures of merit closer to .3-.6. These improvements highlight the fact that in addition to development of novel materials for thermoelectric applications, using different processing techniques to design microstructure is a viable and worthwhile effort. In fact, it often makes sense to work to optimize both composition and microstructure.

Efficiency

The typical efficiency of TEGs is around 5–8%. Older devices used bimetallic junctions and were bulky. More recent devices use highly doped semiconductors made from bismuth telluride (Bi_2Te_3), lead telluride (PbTe), calcium manganese oxide ($Ca_2Mn_3O_8$), or combinations thereof, depending on temperature. These are solid-state devices and unlike dynamos have no moving parts, with the occasional exception of a fan or pump. For a discussion of the factors determining and limiting efficiency, and ongoing efforts to improve the efficiency.

Uses

Thermoelectric generators have a variety of applications. Frequently, thermoelectric generators are used for low power remote applications or where bulkier but more efficient heat engines such as Stirling engines would not be possible. Unlike heat engines, the solid state electrical components typically used to perform thermal to electric energy conversion have no moving parts. The thermal to electric energy conversion can be performed using components that require no maintenance, have inherently high reliability, and can be used to construct generators with long service-free lifetimes. This makes thermoelectric generators well suited for equipment with low to modest power needs in remote uninhabited or inaccessible locations such as mountaintops, the vacuum of space, or the deep ocean.

- Common application is the use of thermoelectric generators on gas pipelines. For example, for cathodic protection, radio communication, and other telemetry. On gas pipelines for power consumption of up to 5 kW thermal generators are preferable to other power sources. The manufacturers of generators for gas pipelines are Gentherm Global Power Technologies (Formerly Global Thermoelectric), (Calgary, Canada) and TELGEN (Russia).

- Thermoelectric Generators are primarily used as remote and off-grid power generators for unmanned sites. They are the most reliable power generator in such situations as they do not have moving parts (thus virtually maintenance free), work day and night, perform under all weather conditions, and can work without battery backup. Although Solar Photovoltaic systems are also implemented in remote sites, Solar PV may not be a suitable solution where solar radiation is low, i.e. areas at higher latitudes with snow or no sunshine, areas with lots of cloud or tree canopy cover, dusty deserts, forests, etc.

- Gentherm Global Power Technologies (GPT) formerly known as Global Thermoelectric (Canada) has Hybrid Solar-TEG solutions where the Thermoelectric Generator backs up the Solar-PV, such that if the Solar panel is down and the backup battery backup goes into deep discharge then a sensor starts the TEG as a backup power source until the Solar is up again. The TEG heat can be produced by a low pressure flame fueled by Propane or Natural Gas.

- Many space probes, including the Mars *Curiosity* rover, generate electricity using a radioisotope thermoelectric generator whose heat source is a radioactive element.

- Cars and other automobiles produce waste heat (in the exhaust and in the cooling agents). Harvesting that heat energy, using a thermoelectric generator, can increase the fuel efficiency of the car.

- In addition to automobiles, waste heat is also generated in many other places, such as in industrial processes and in heating (wood stoves, outdoor boilers, cooking, oil and gas fields, pipelines, and remote communication towers).

- Microprocessors generate waste heat. Researchers have considered whether some of that energy could be recycled.

- Solar cells use only the high frequency part of the radiation, while the low frequency heat energy is wasted. Several patents about the use of thermoelectric devices in tandem with solar cells have been filed. The idea is to increase the efficiency of the combined solar/thermoelectric system to convert the solar radiation into useful electricity.

- The Maritime Applied Physics Corporation in Baltimore, Maryland is developing a thermoelectric generator to produce electric power on the deep-ocean offshore seabed using the temperature difference between cold seawater and hot fluids released by hydrothermal vents, hot seeps, or from drilled geothermal wells. A high reliability source of seafloor electric power is needed for ocean observatories and sensors used in the geological, environmental, and ocean sciences, by seafloor mineral and energy resource developers, and by the military.

- Ann Makosinski from British Columbia, Canada has developed several devices using Peltier tiles to harvest heat (from a human hand, the forehead, and hot beverage) that claims to generate enough electricity to power an LED light or charge a mobile device, although the inventor admits that the brightness of the LED light is not competitive with those on the market.

Practical Limitations

Besides low efficiency and relatively high cost, practical problems exist in using thermoelectric devices in certain types of applications resulting from a relatively high electrical output resistance, which increases self-heating, and a relatively low thermal conductivity, which makes them unsuitable for applications where heat removal is critical, as with heat removal from an electrical device such as microprocessors.

- High generator output resistance: In order to get voltage output levels in the range required by digital electrical devices, a common approach is to place many thermoelectric elements in series within a generator module. The element's voltages add, but so do their individual output resistance. The maximum power transfer theorem dictates that maximum power is delivered to a load when the source and load resistances are identically matched. For low impedance loads near zero ohms, as the generator resistance rises the power delivered to the load decreases. To lower the output resistance, some commercial devices place more individual elements in parallel and fewer in series and employ a boost regulator to raise the voltage to the voltage needed by the load.

- Low thermal conductivity: Because a very high thermal conductivity is required to transport thermal energy away from a heat source such as a digital microprocessor, the low thermal conductivity of thermoelectric generators makes them unsuitable to recover the heat.

- Cold-side heat removal with air: In air-cooled thermoelectric applications, such as when harvesting thermal energy from a motor vehicle's crankcase, the large amount of thermal energy that must be dissipated into ambient air presents a significant challenge. As a thermoelectric generator's cool side temperature rises, the device's differential working temperature decreases. As the temperature rises, the device's electrical resistance increases causing greater parasitic generator self-heating. In motor vehicle applications a supplementary radiator is sometimes used for improved heat removal, though the use of an electric water pump to circulate a coolant adds an additional parasitic loss to total generator output power. Water cooling the thermoelectric generator's cold side, as when generating thermoelectric power from the hot crank case of an inboard boat motor, would not suffer from this disadvantage. Water is a far easier coolant to use effectively in contrast to air.

Future Market

While TEG technology has been used in military and aerospace applications for decades, new TE materials and systems are being developed to generate power using low or high temperatures waste heat, and that could provide a significant opportunity in the near future. These systems can also be scalable to any size and have lower operation and maintenance cost.

In general, investing in TEG technology is increasing rapidly. The global market for thermoelectric generators is estimated to be US\$320 million in 2015. A recent study estimated that TEG is expected to reach \$720 million in 2021 with a growth rate of 14.5%. Today, North America capture 66% of the market share and it will continue to be the biggest market in the near future. However, Asia-Pacific and European countries are projected to grow at relatively higher rates. A study found that the Asia-Pacific market would grow at a Compound Annual Growth Rate (CAGR) of 18.3% in the period from 2015 to 2020 due to the high demand of thermoelectric generators by the automotive industries to increase overall fuel efficiency, as well as the growing industrialization in the region.

Low power TEG or "Sub-watt" (i.e. generating up to 1 Watt peak) market is a growing part of the TEG market, capitalizing on latest technologies. Main applications are sensors, low power applications and more globally Internet of things applications. A specialized market research company indicated that 100 000 units have been shipped in 2014 and expects 9 million units per year by 2020.

Thermocouples are the most popular temperature sensors. They are inexpensive, interchangeable, have standard connectors and can measure a wide range of temperatures. Their main limitation is accuracy as the system errors of less than 1°C can be difficult to achieve. Following figure represents internal construction of thermocouple and its circuitry.

A thermocouple is constructed of two dissimilar metal wires joined at one end. It works on the principle of "Seabeck effect" whereby electromagnetic force is generated when two dissimilar metals are joined at two different temperature ends. When one end of each wire is connected to a measuring instrument, the thermocouple becomes a sensitive and highly accurate measuring device. Heating the measuring junction of the thermocouple produces a voltage which is greater than the voltage across the reference junction. The difference between the two voltages is proportional to the difference in temperature and can be measured on the voltmeter (in mV). Thermocouples may be constructed of several different combinations of materials. The most important factor to be considered when selecting a pair of materials is the "thermoelectric difference" between the two materials. A significant difference between the two materials will result in better thermocouple performance.

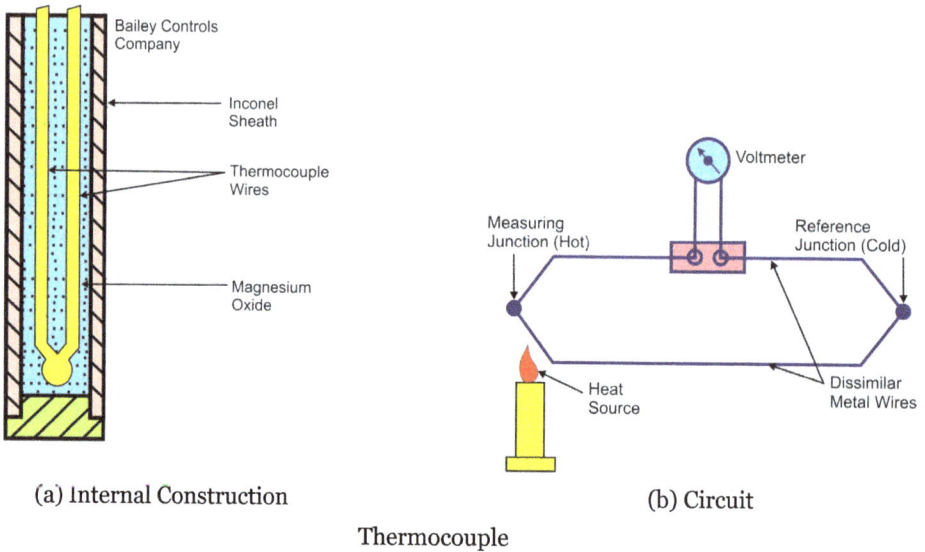

(a) Internal Construction   (b) Circuit

Thermocouple

Resistance Thermometer

Resistance thermometers, also called resistance temperature detectors (RTDs), are sensors used to measure temperature. Many RTD elements consist of a length of fine wire wrapped around a ceramic or glass core but other constructions are also used. The RTD wire is a pure material, typically platinum, nickel, or copper. The material has an accurate resistance/temperature relationship which is used to provide an indication of temperature. As RTD elements are fragile, they are often housed in protective probes.

RTDs, which have higher accuracy and repeatability, are slowly replacing thermocouples in industrial applications below 600 °C.

Resistance/Temperature Relationship of Metals

Common RTD sensing elements constructed of platinum, copper or nickel have a repeatable resistance versus temperature relationship (R vs T) and operating temperature range. The R vs T relationship is defined as the amount of resistance change of the sensor per degree of temperature change. The relative change in resistance (temperature coefficient of resistance) varies only slightly over the useful range of the sensor.

Platinum was proposed by Sir William Siemens as an element for a resistance temperature detector at the Bakerian lecture in 1871: it is a noble metal and has the most stable resistance–temperature relationship over the largest temperature range. Nickel elements have a limited temperature range because the amount of change in resistance per degree of change in temperature becomes very non-linear at temperatures over 572 °F (300 °C). Copper has a very linear resistance–temperature relationship; however, copper oxidizes at moderate temperatures and cannot be used over 302 °F (150 °C).

Platinum is the best metal for RTDs due to its very linear resistance–temperature relationship, highly repeatable over a wide temperature range. The unique properties of platinum make it the material of choice for temperature standards over the range of −272.5 °C to 961.78 °C. It is used in the sensors that define the International Temperature Standard, ITS-90. Platinum is chosen also because of its chemical inertness.

The significant characteristic of metals used as resistive elements is the linear approximation of the resistance versus temperature relationship between 0 and 100 °C. This temperature coefficient of resistance is denoted by α and is usually given in units of $\Omega/(\Omega \cdot °C)$:

$$\alpha = \frac{R_{100} - R_0}{100°C \cdot R_0},$$

where

R_0 is the resistance of the sensor at 0 °C,

R_{100} is the resistance of the sensor at 100 °C.

Pure platinum has $\alpha = 0.003925$ $\Omega/(\Omega \cdot °C)$ in the 0 to 100 °C range and is used in the construction of laboratory-grade RTDs. Conversely, two widely recognized standards for industrial RTDs IEC 60751 and ASTM E-1137 specify $\alpha = 0.00385$ $\Omega/(\Omega \cdot °C)$. Before these standards were widely adopted, several different α values were used. It is still possible to find older probes that are made with platinum that have $\alpha = 0.003916$ $\Omega/(\Omega \cdot °C)$ and 0.003902 $\Omega/(\Omega \cdot °C)$.

These different α values for platinum are achieved by doping; basically, carefully introducing impurities into the platinum. The impurities introduced during doping become embedded in the lattice structure of the platinum and result in a different R vs. T curve and hence α value.

Calibration

To characterize the R vs T relationship of any RTD over a temperature range that represents the planned range of use, calibration must be performed at temperatures other than 0 °C and 100 °C. This is necessary to meet calibration requirements. Although RTDs are considered to be linear in operation, it must be proven that they are accurate with regard to the temperatures with which they will actually be used. Two common calibration methods are the fixed-point method and the comparison method.

Fixed Point Calibration

is used for the highest-accuracy calibrations by national metrology laboratories. It uses the triple point, freezing point or melting point of pure substances such as water, zinc, tin, and argon to generate a known and repeatable temperature. These cells allow the user to reproduce actual conditions of the ITS-90 temperature scale. Fixed-point calibrations provide extremely accurate calibrations (within ±0.001 °C). A common fixed-point calibration method for industrial-grade probes is the ice bath. The equipment is inexpensive, easy to use, and can accommodate several sensors at once. The ice point is designated as a secondary standard because its accuracy is ±0.005 °C (±0.009 °F), compared to ±0.001 °C (±0.0018 °F) for primary fixed points.

Comparison Calibrations

is commonly used with secondary SPRTs and industrial RTDs. The thermometers being calibrated are compared to calibrated thermometers by means of a bath whose temperature is uniformly stable. Unlike fixed-point calibrations, comparisons can be made at any temperature between −100 °C and 500 °C (−148 °F to 932 °F). This method might be more cost-effective, since several sensors can be calibrated simultaneously with automated equipment. These electrically heated and well-stirred baths use silicone oils and molten salts as the medium for the various calibration temperatures.

Element Types

There are five main categories of RTD sensors: thin-film, wire-wound, and coiled elements. While these types are the ones most widely used in industry, there are some places where other more exotic shapes are used, for example carbon resistors are used at ultra-low temperatures (−173 °C to −273 °C).

Carbon Resistor Elements

are cheap and widely used. They have very reproducible results at low temperatures. They are the most reliable form at extremely low temperatures. They generally do not suffer from significant hysteresis or strain gauge effects.

Strain-free Elements

use a wire coil minimally supported within a sealed housing filled with an inert gas. These sensors work up to 961.78 °C and are used in the SPRTs that define ITS-90. They consist of platinum wire loosely coiled over a support structure, so the element is free to expand and contract with temperature. They are very susceptible to shock and vibration, as the loops of platinum can sway back and forth, causing deformation.

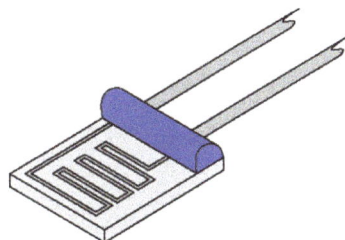

Thin-film PRT

Thin-film Elements

have a sensing element that is formed by depositing a very thin layer of resistive material, normally platinum, on a ceramic substrate (plating). This layer is usually just 10 to 100 ångströms (1 to 10 nanometers) thick. This film is then coated with an epoxy or glass that helps protect the deposited film and also acts as a strain relief for the external lead wires. Disadvantages of this type are that they are not as stable as their wire-wound or coiled counterparts. They also can only be used over a limited temperature range due to the different expansion rates of the substrate and resistive deposited giving a "strain gauge" effect that can be seen in the resistive temperature coefficient. These elements work with temperatures to 300 °C (572 °F) without further packaging, but can operate up to 600 °C (1,112 °F) when suitably encapsulated in glass or ceramic. Nowadays there are special high-temperature RTD elements that can be used up to 900 °C (1,652 °F) with the right encapsulation.

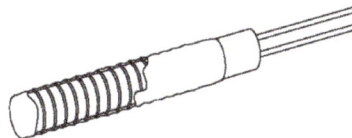

Wire-wound PRT

Wire-wound Elements

can have greater accuracy, especially for wide temperature ranges. The coil diameter provides a compromise between mechanical stability and allowing expansion of the wire to minimize strain and consequential drift. The sensing wire is wrapped around an insulating mandrel or core. The winding core can be round or flat, but must be an electrical insulator. The coefficient of thermal expansion of the winding core material is matched to the sensing wire to minimize any mechanical strain. This strain on the element wire will result in a thermal measurement error. The sensing wire is connected to a larger wire, usually referred to as the element lead or wire. This wire is selected to be compatible with the sensing wire, so that the combination does not generate an emf that would distort the thermal measurement. These elements work with temperatures to 660 °C.

Coil-element PRT

Coiled Elements

have largely replaced wire-wound elements in industry. This design has a wire coil that can expand freely over temperature, held in place by some mechanical support, which lets the coil keep its shape. This "strain free" design allows the sensing wire to expand and contract free of influence from other materials; in this respect it is similar to the SPRT, the primary standard upon which ITS-90 is based, while providing the durability necessary for industrial use. The basis of the sensing element is a small coil of platinum sensing wire. This coil resembles a filament in an incandescent light bulb. The housing or mandrel is a hard fired ceramic oxide tube with equally spaced bores that run transverse to the axes. The coil is inserted in the bores of the mandrel and then packed with a very finely ground ceramic powder. This permits the sensing wire to move, while still remaining in good thermal contact with the process. These elements work with temperatures to 850 °C.

The current international standard that specifies tolerance and the temperature-to-electrical resistance relationship for platinum resistance thermometers (PRTs) is IEC 60751:2008; ASTM E1137 is also used in the United States. By far the most common devices used in industry have a nominal resistance of 100 ohms at 0 °C and are called Pt100 sensors ("Pt" is the symbol for platinum, "100" for the resistance in ohms at 0 °C). It is also possible to get Pt1000 sensors, where 1000 is for the resistance in ohms at 0 °C. The sensitivity of a standard 100 Ω sensor is a nominal 0.385 Ω/°C. RTDs with a sensitivity of 0.375 and 0.392 Ω/°C, as well as a variety of others, are also available.

Function

Resistance thermometers are constructed in a number of forms and offer greater stability, accuracy and repeatability in some cases than thermocouples. While thermocouples use the Seebeck effect to generate a voltage, resistance thermometers use electrical resistance and require a power source to operate. The resistance ideally varies nearly linearly with temperature per the Callendar-Van Dusen equation.

The platinum detecting wire needs to be kept free of contamination to remain stable. A platinum wire or film is supported on a former in such a way that it gets minimal differential expansion or other strains from its former, yet is reasonably resistant to vibration. RTD assemblies made from iron or copper are also used in some applications. Commercial platinum grades exhibit a temperature coefficient of resistance 0.00385/°C (0.385%/°C) (European Fundamental Interval). The sensor is usually made to have a resistance of 100 Ω at 0 °C. This is defined in BS EN 60751:1996 (taken from IEC 60751:1995). The American Fundamental Interval is 0.00392/°C, based on using a purer grade of platinum than the European standard. The American standard is from the Scientific Apparatus Manufacturers Association (SAMA), who are no longer in this standards field. As a result, the "American standard" is hardly the standard even in the US.

Lead-wire resistance can also be a factor; adopting three- and four-wire, instead of two-wire, connections can eliminate connection-lead resistance effects from measurements; three-wire connection is sufficient for most purposes and is an almost universal industrial practice. Four-wire connections are used for the most precise applications.

Advantages and Limitations

The advantages of platinum resistance thermometers include:

- High accuracy

- Low drift

- Wide operating range

- Suitability for precision applications.

Limitations: RTDs in industrial applications are rarely used above 660 °C. At temperatures above 660 °C it becomes increasingly difficult to prevent the platinum from becoming contaminated by impurities from the metal sheath of the thermometer. This is why laboratory standard thermometers replace the metal sheath with a glass construction. At very low temperatures, say below −270 °C (3 K), because there are very few phonons, the resistance of an RTD is mainly determined by impurities and boundary scattering and thus basically independent of temperature. As a result, the sensitivity of the RTD is essentially zero and therefore not useful.

Compared to thermistors, platinum RTDs are less sensitive to small temperature changes and have a slower response time. However, thermistors have a smaller temperature range and stability.

RTDs vs Thermocouples

The two most common ways of measuring industrial temperatures are with resistance temperature detectors (RTDs) and thermocouples. Choice between them is usually determined by four factors.

Temperature

If process temperatures are between −200 and 500 °C (−328.0 and 932.0 °F), an industrial RTD is the preferred option. Thermocouples have a range of −180 to 2,320 °C (−292.0 to 4,208.0 °F), so for temperatures above 500 °C (932 °F) they are the only contact temperature measurement device.

Response Time

If the process requires a very fast response to temperature changes (fractions of a second as opposed to seconds), then a thermocouple is the best choice. Time response is measured by immersing the sensor in water moving at 1 m/s (3 ft/s) with a 63.2% step change.

Size

A standard RTD sheath is 3.175 to 6.35 mm (0.1250 to 0.2500 in) in diameter; sheath diameters for thermocouples can be less than 1.6 mm (0.063 in).

Accuracy and Stability Requirements

If a tolerance of 2 °C is acceptable and the highest level of repeatability is not required, a thermocouple will serve. RTDs are capable of higher accuracy and can maintain stability for many years, while thermocouples can drift within the first few hours of use.

Construction

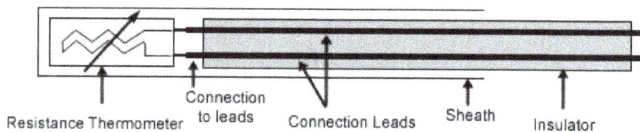

Resistance Thermometer   Connection to leads   Connection Leads   Sheath   Insulator

These elements nearly always require insulated leads attached. PVC, silicone rubber or PTFE insulators are used at temperatures below about 250 °C. Above this, glass fibre or ceramic are used. The measuring point, and usually most of the leads, require a housing or protective sleeve, often made of a metal alloy that is chemically inert to the process being monitored. Selecting and designing protection sheaths can require more care than the actual sensor, as the sheath must withstand chemical or physical attack and provide convenient attachment points.

Wiring Configurations

Two-wire Configuration

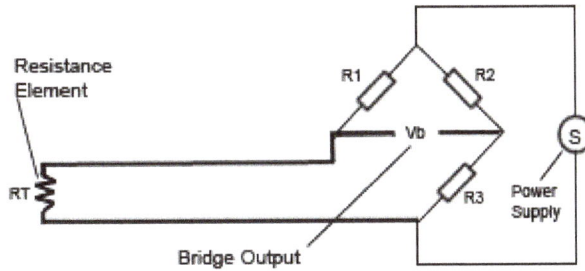

The simplest resistance-thermometer configuration uses two wires. It is only used when high accuracy is not required, as the resistance of the connecting wires is added to that of the sensor, leading to errors of measurement. This configuration allows use of 100 meters of cable. This applies equally to balanced bridge and fixed bridge system.

For a balanced bridge usual setting is with R2 = R3, and R1 around the middle of the range of the RTD. So for example, if we are going to measure between 0 and 100 °C (32 and 212 °F), RTD resistance will range from 100 Ω to 138.5 Ω. We would choose R1 = 120 Ω. In that way we get a small measured voltage in the bridge.

Three-wire Configuration

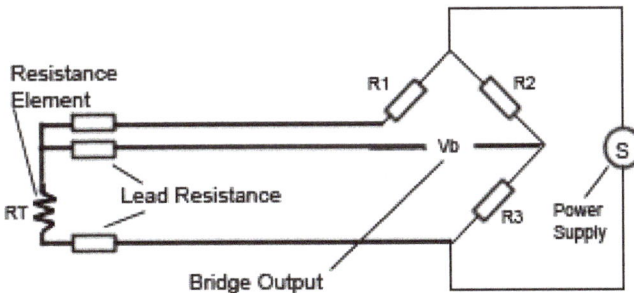

In order to minimize the effects of the lead resistances, a three-wire configuration can be used. In this method the two leads to the sensor are on adjoining arms. There is a lead resistance in each arm of the bridge, so that the resistance is cancelled out if the two lead resistances are accurately the same. This configuration allows up to 600 metres (2,000 feet) of cable.

As in the case with the 2-wire connection, the usual setting is with R2 = R3, and R1 around the middle of the range of the RTD.

Four-wire Configuration

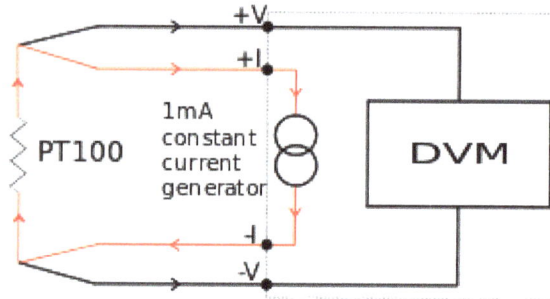

The four-wire resistance configuration increases the accuracy of measurement of resistance. Four-terminal sensing eliminates voltage drop in the measuring leads as a contribution to error. To increase accuracy further, any residual thermoelectric voltages generated by different wire types or screwed connections are eliminated by reversal of the direction of the 1 mA current and the leads to the DVM (digital voltmeter). The thermoelectric voltages will be produced in one direction only. By averaging the reversed measurements, the thermoelectric error voltages are cancelled out.

Classifications of RTDs

The highest-accuracy of all PRTs are the *Standard Platinum Resistance Thermometers* (SPRTs). This accuracy is achieved at the expense of durability and cost. The SPRT elements are wound from reference-grade platinum wire. Internal lead wires are usually made from platinum, while internal supports are made from quartz or fused silica. The sheaths are usually made from quartz or sometimes Inconel, depending on temperature range. Larger-diameter platinum wire is used, which drives up the cost and results in a lower resistance for the probe (typically 25.5 Ω). SPRTs have a wide temperature range (−200 °C to 1000 °C) and are approximately accurate to ±0.001 °C over the temperature range. SPRTs are only appropriate for laboratory use.

Another classification of laboratory PRTs is *Secondary-Standard Platinum Resistance Thermometers* (Secondary SPRTs). They are constructed like the SPRT, but the materials are more cost-effective. SPRTs commonly use reference-grade, high-purity smaller-diameter platinum wire, metal sheaths and ceramic type insulators. Internal lead wires are usually a nickel-based alloy. Secondary SPRTs are more limited in temperature range (−200 °C to 500 °C) and are approximately accurate to ±0.03 °C over the temperature range.

Industrial PRTs are designed to withstand industrial environments. They can be almost as durable as a thermocouple. Depending on the application, industrial PRTs can use thin-film or coil-wound elements. The internal lead wires can range from PTFE-insulated stranded nickel-plated copper to silver wire, depending on the sensor size and application. Sheath material is typically stainless steel; higher-temperature applications may demand Inconel. Other materials are used for specialized applications.

History

The application of the tendency of electrical conductors to increase their electrical resistance with rising temperature was first described by Sir William Siemens at the Bakerian Lecture of 1871 before the Royal Society of Great Britain. The necessary methods of construction were established by Callendar, Griffiths, Holborn and Wein between 1885 and 1900.

Standard Resistance Thermometer Data

Temperature sensors are usually supplied with thin-film elements. The resistance elements are rated in accordance with BS EN 60751:2008 as:

Tolerance class	Valid range
F 0.3	−50 to +500 °C
F 0.15	−30 to +300 °C
F 0.1	0 to +150 °C

Resistance-thermometer elements functioning up to 1000 °C can be supplied. The relation between temperature and resistance is given by the Callendar-Van Dusen equation:

$$R_T = R_0 \left[1 + AT + BT^2 + CT^3 \left(T - 100 \right) \right] \left(-200^\circ C < T < 0^\circ C \right),$$

$$R_T = R_0 \left[1 + AT + BT^2 \right] \left(0^\circ C \leq T < 850^\circ C \right).$$

Here R_T is the resistance at temperature T, R_0 is the resistance at 0 °C, and the constants (for an $\alpha = 0.00385$ platinum RTD) are:

$$A = 3.9083 \times 10^{-3} \, ^\circ C^{-1},$$

$$B = -5.775 \times 10^{-7} \, ^\circ C^{-2},$$

$$C = -4.183 \times 10^{-12} \, ^\circ C^{-4}.$$

Since the B and C coefficients are relatively small, the resistance changes almost linearly with the temperature.

For positive temperature, solution of the quadratic equation yields the following relationship between temperature and resistance:

$$T = \frac{-A+\sqrt{A^2-4B\left(1-\dfrac{R_T}{R_0}\right)}}{2B}.$$

Then for a four-wire configuration with a 1 mA precision current source the relationship between temperature and measured voltage V_T is

$$T = \frac{-A+\sqrt{A^2-40B\left(0.1-V_T\right)}}{2B}.$$

Temperature-dependent Resistances for Various Popular Resistance Thermometers

Tem-perature in °C	ITS-90 Pt100	Pt100 Typ: 404	Pt1000 Typ: 501	PTC Typ: 201	NTC Typ: 101	NTC Typ: 102	NTC Typ: 103	NTC Typ: 104	NTC Typ: 105
				Resistance in Ω					
−50	79.901192	80.31	803.1	1032					
−45	81.925089	82.29	822.9	1084					
−40	83.945642	84.27	842.7	1135			50475		
−35	85.962913	86.25	862.5	1191			36405		
−30	87.976963	88.22	882.2	1246			26550		
−25	89.987844	90.19	901.9	1306		26083	19560		
−20	91.995602	92.16	921.6	1366		19414	14560		
−15	94.000276	94.12	941.2	1430		14596	10943		
−10	96.001893	96.09	960.9	1493		11066	8299		
−5	98.000470	98.04	980.4	1561	31389	8466			
0	99.996012	100.00	1000.0	1628	23868	6536			
5	101.988430	101.95	1019.5	1700	18299	5078			
10	103.977803	103.90	1039.0	1771	14130	3986			
15	105.964137	105.85	1058.5	1847	10998				
20	107.947437	107.79	1077.9	1922	8618				
25	109.927708	109.73	1097.3	2000	6800			15000	
30	111.904954	111.67	1116.7	2080	5401			11933	
35	113.879179	113.61	1136.1	2162	4317			9522	
40	115.850387	115.54	1155.4	2244	3471			7657	
45	117.818581	117.47	1174.7	2330				6194	
50	119.783766	119.40	1194.0	2415				5039	
55	121.745943	121.32	1213.2	2505				4299	27475
60	123.705116	123.24	1232.4	2595				3756	22590

65	125.661289	125.16	1251.6	2689					18668
70	127.614463	127.07	1270.7	2782					15052
75	129.564642	128.98	1289.8	2880					12932
80	131.511828	130.89	1308.9	2977					10837
85	133.456024	132.80	1328.0	3079					9121
90	135.397232	134.70	1347.0	3180					7708
95	137.335456	136.60	1366.0	3285					6539
100	139.270697	138.50	1385.0	3390					
105	141.202958	140.39	1403.9						
110	143.132242	142.29	1422.9						
150	158.459633	157.31	1573.1						
200	177.353177	175.84	1758.4						

Resistance Temperature Detector (RTD)

The RTD incorporates pure metals or certain alloys that increase in resistance as temperature increases and, conversely, decrease in resistance as temperature decreases. RTDs act somewhat like an electrical transducer, converting changes in temperature to voltage signals by the measurement of resistance. The metals that are best suited for use as RTD sensors are pure, of uniform quality, stable within a given range of temperature, and able to give reproducible resistance-temperature readings. RTD elements are normally constructed of platinum, copper, or nickel. These metals are best suited for RTD applications because of their linear resistance-temperature characteristics.

$$R = R_0 \left(1 + \propto T\right) \qquad\qquad (VII.6)$$

The above equation represents the Resistance vs. Temperature relationship where R is the resistance and temperature T , with reference Resistance R_0 and coefficient of resistance as $\propto$. The coefficient of resistance is the change in resistance per degree change in temperature, usually expressed as a percentage per degree of temperature.

Flow Measuring Devices

Various flow measurement devices are already taught in the relevant course of Fluid Mechanics such as Orificemeter, Venturimeter, pitot tube, rotameter etc . A few other types of flow meters are discussed here.

Hot-wire Anemometer

The hot-wire anemometer, consists of an electrically heated, fine platinum wire which is immersed into the flow. As the fluid velocity increases, the rate of heat flow from the heated wire to the flow stream increases. Thus, a cooling effect on the wire electrode occurs, causing its electrical resistance to change.

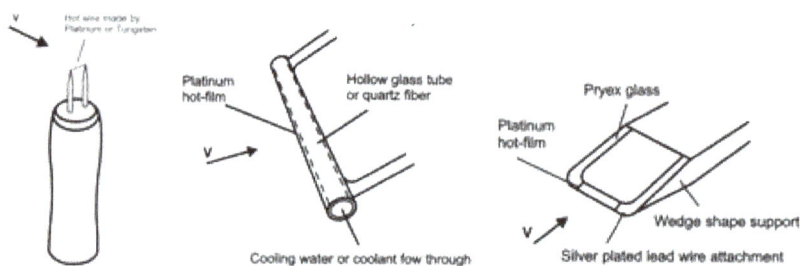

Schematic of a Hot-wire anemometer

In a constant-current anemometer, the fluid velocity is determined from a measurement of the resulting change in wire resistance. In a constant-resistance anemometer, fluid velocity is determined from the current needed to maintain a constant wire temperature and, thus, the resistance constant. Typically, the anemometer wire is made of platinum or tungsten and is 4 μ in diameter and 1 mm in length. Typical commercially available hot-wire anemometers have a flat frequency (< 3 dB) up to 17 kHz at the average velocity of 30 ft/s, 30 kHz at 100 ft/s, or 50 kHz at 300 ft/s. Due to the tiny size of the wire, it is fragile and thus suitable only for clean fluid. In liquid flow or rugged gas flow, a platinum hot-film coated on a 25 ~ 150 mm diameter quartz fiber or hollow glass tube can be used instead, as shown in the schematic. Another alternative is a pyrex glass wedge coated with a thin platinum hot-film edge tip, as shown schematically.

Nutating Disc Displacement Meter

In a displacement flowmeter, all of the fluid passes through the meter in almost completely isolated quantities. The number of these quantities is counted and indicated in terms of volume or weight units by a register. The most common type of displacement flowmeter is the nutating disk, or wobble plate meter. A typical nutating disk is shown in Figure.

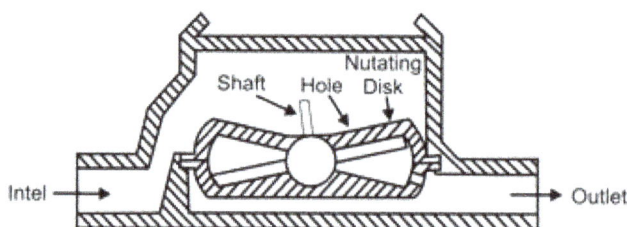

Schematic of a Nutating disk displacement meter

The movable element is a circular disk which is attached to a central ball. A shaft is fastened to the ball and held in an inclined position by a cam or roller. The disk is mounted in a chamber which has spherical side walls and conical top and bottom surfaces. The fluid enters an opening in the spherical wall on one side of the partition and leaves through the other side. As the fluid flows through the chamber, the disk wobbles, or executes a nutating motion. Since the volume of fluid required to make the disc complete one revolution is known, the total flow through a nutating disc can be calculated by multiplying the number of disc rotations by the known volume of fluid. To measure this flow, the motion of the shaft generates a cone with the point, or apex, down. The top of the shaft operates a revolution counter, through a crank and set of gears, which is calibrated to indicate total system flow.

Level Measuring Devices

The differential pressure (P) detector method is also used for liquid level measurement which uses a P detector connected to the bottom of the tank being monitored. The higher pressure, caused by the fluid in the tank, is compared to a lower reference pressure (usually atmospheric). This comparison takes place in the P detector. In addition to this a couple of other level detectors are discussed below.

Ball Float

The ball float method is a direct reading liquid level mechanism. The most practical design for the float is a hollow metal ball or sphere. However, there are no restrictions to the size, shape, or material used. The design consists of a ball float attached to a rod, which in turn is connected to a rotating shaft which indicates level on a calibrated scale.

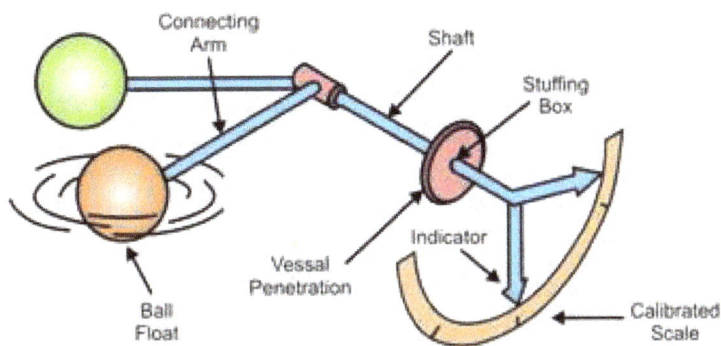

Schematic of a ball float level mechanism

The operation of the ball float is simple. The ball floats on top of the liquid in the tank. If the liquid level changes, the float will follow and change the position of the pointer attached to the rotating shaft.

Magnetic Bond Level Indicator

The magnetic bond mechanism consists of a magnetic float which rises and falls with changes in level. The float travels outside of an on-magnetic tube which houses an inner magnet connected to a level indicator.

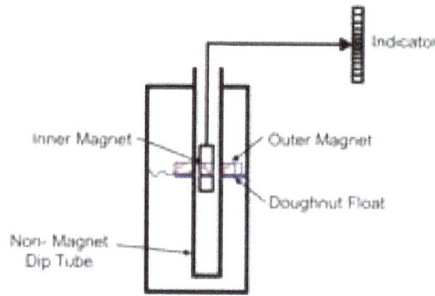

Schematic of a magnetic bond level indicator

When the float rises and falls, the outer magnet will attract the inner magnet, causing the inner magnet to follow the level within the vessel.

References

- Katz, Eric; Light, Andrew; Thompson, William (2002). Controlling technology : contemporary issues (2nd ed.). Amherst, NY: Prometheus Books. ISBN 978-1573929837. Retrieved 9 March 2016

- Wildhack, W. A. (22 October 1954). "Instrumentation--Revolution in Industry, Science, and Warfare". Science. 120 (3121): 15A–15A. doi:10.1126/science.120.3121.15A. Retrieved 9 March 2016

- Baird, D. (1993). "Analytical chemistry and the 'big' scientific instrumentation revolution". Annals of Science. 50: 267–290. doi:10.1080/00033799300200221.

- Reinhardt, Carsten, ed. (2001). Chemical sciences in twentieth century (1st ed.). Weinheim: Wiley-VCH. ISBN 978-3527302710

- Snyder, G. (Oct 2003). "Thermoelectric Efficiency and Compatibility". Physical Review Letters. 91 (14). doi:10.1103/physrevlett.91.148301

- Kerlin, T.W. & Johnson, M.P. (2012). Practical Thermocouple Thermometry (2nd Ed.). Research Triangle Park: ISA. pp. 110–112. ISBN 978-1-937560-27-0

- Kim, D.S. (January 2008). "Solar refrigeration options – a state-of-the-art review". International Journal of Refrigeration. 31 (1): 3–15. doi:10.1016/j.ijrefrig.2007.07.011. Retrieved 8 November 2016

- Flammable Vapor Ignition Resistant Water Heaters: Service Manual (238-44943-00D) (PDF). Bradford White. pp. 11–16. Retrieved 11 June 2014

- Kraemer, D; Hu, L; Muto, A; Chen, X; Chen, G; Chiesa, M (2008), "Photovoltaic-thermoelectric hybrid systems: A general optimization methodology", Applied Physics Letters, 92: 243503, doi:10.1063/1.2947591

- Hablanian, M. H. (1997) High-Vacuum Technology: A Practical Guide, Second Ed., Marcel Dekker Inc., pp. 19–22, 45–47 & 438–443, ISBN 0-8247-9834-1

- Cojocaru-Mirédin, Oana. "Thermoelectric Materials Design by controlling the microstructure and composition". Max-Planck Institut. Retrieved 8 November 2016

- "Harvesting Wasted Heat in a Microprocessor Using Thermoelectric Generators: Modeling, Analysis and Measurement". 2008 Design, Automation and Test in Europe. doi:10.1109/DATE.2008.4484669

Permissions

All chapters in this book are published with permission under the Creative Commons Attribution Share Alike License or equivalent. Every chapter published in this book has been scrutinized by our experts. Their significance has been extensively debated. The topics covered herein carry significant information for a comprehensive understanding. They may even be implemented as practical applications or may be referred to as a beginning point for further studies.

We would like to thank the editorial team for lending their expertise to make the book truly unique. They have played a crucial role in the development of this book. Without their invaluable contributions this book wouldn't have been possible. They have made vital efforts to compile up to date information on the varied aspects of this subject to make this book a valuable addition to the collection of many professionals and students.

This book was conceptualized with the vision of imparting up-to-date and integrated information in this field. To ensure the same, a matchless editorial board was set up. Every individual on the board went through rigorous rounds of assessment to prove their worth. After which they invested a large part of their time researching and compiling the most relevant data for our readers.

The editorial board has been involved in producing this book since its inception. They have spent rigorous hours researching and exploring the diverse topics which have resulted in the successful publishing of this book. They have passed on their knowledge of decades through this book. To expedite this challenging task, the publisher supported the team at every step. A small team of assistant editors was also appointed to further simplify the editing procedure and attain best results for the readers.

Apart from the editorial board, the designing team has also invested a significant amount of their time in understanding the subject and creating the most relevant covers. They scrutinized every image to scout for the most suitable representation of the subject and create an appropriate cover for the book.

The publishing team has been an ardent support to the editorial, designing and production team. Their endless efforts to recruit the best for this project, has resulted in the accomplishment of this book. They are a veteran in the field of academics and their pool of knowledge is as vast as their experience in printing. Their expertise and guidance has proved useful at every step. Their uncompromising quality standards have made this book an exceptional effort. Their encouragement from time to time has been an inspiration for everyone.

The publisher and the editorial board hope that this book will prove to be a valuable piece of knowledge for students, practitioners and scholars across the globe.

Index

www.ingramcontent.com/pod-product-compliance
Lightning Source LLC
Chambersburg PA
CBHW062003190326
41458CB00009B/2955